NDE-Vol. 21

# NDE CHALLENGES OF THE 21ST CENTURY: THEORY TO PRACTICE

**presented at**

THE 2001 ASME MECHANICAL ENGINEERING CONGRESS AND EXPOSITION
NOVEMBER 11–16, 2001
NEW YORK, NEW YORK

**sponsored by**

THE NONDESTRUCTIVE EVALUATION ENGINEERING DIVISION, ASME

**edited by**

CETIN CETINKAYA
CLARKSON UNIVERSITY

THE AMERICAN SOCIETY OF MECHANICAL ENGINEERS
Three Park Avenue / New York, N.Y. 10016

ISBN No. 0-7918-3583-9

## FOREWORD

The general theme for the ASME Nondestructive Evaluation Engineering Division's Symposium for the 2001 ASME International Mechanical Engineering Congress and Exposition is *"NDE Challenges of the 21st Century: Theory to Practice"*. The focus in the 2001 sessions are on NDE applications with
(*i*) the new trends in shrinking the dimensions of (micro-to-nano) structures, and
(*ii*) the aging structures.

The reduction in the dimensions of structures has started posing new challenges to the NDE communities worldwide. It appears that the engineering applications of the 21st century will not only require the adoption of micro-scale structures (such as MEMS and ICs), but also the use of nano-scale structures in engineering design may become a routine practice. We see a lot of potential and opportunities for the NDE community in this emerging area.

The infrastructure in the industrialized world has been showing its age for over one decade. The reliable assessment of the state of an aging structure of various dimensions has become a critical economic issue. It is reasonable to expect that this requirement will become more demanding going into the 21st century, as the number of aging structures increases. The 2001 NDE sessions will provide a forum to identify, discuss and address the challenges in these exciting areas of engineering and research.

The topics of the papers presented at the Symposium on a diverse spectrum of NDE and NDT fields: Ultrasonic Techniques, Aging Aircraft and Infrastructure, NDE of MEMS and microstructures, Laser Ultrasonics, Material Characterization, NDE of Composites, Thermal Methods, Fatigue, Corrosion, Damage Detection and Sizing, Bonds, Interfaces and Coating, and Emerging Techniques.

Contributed manuscripts were peer-reviewed. On behalf of the NDE Division, I would like to acknowledge the follower reviewers for their feedback and comments on the content of the submissions:
T. Kundu of University of Arizona, B.M. Liaw of The City College of The City University of New York, N. Meyendorf of University of Dayton (Center for Materials Diagnostics) and M. L. Peterson of University of Maine.

I hope that you will find the coverage of topics helpful and interesting, and continue to participate in NDE activities in ASME.

Cetin Cetinkaya
Editor
Dept. of Mechanical and Aeronautical Engineering
Clarkson University

## CONTENTS


**CIVIL AND LARGE STRUCTURES**

Ultrasonic Interfacial Waves in a Finitely Strained Elastic Structure
*D. A. Sotiropoulos* .......... 1

Use of Lamb Waves To Predict Current Damaged State of Mortar
*Michael S. Keller, Tribikram Kundu, and Chandrakant S. Desai* .......... 9

Quantitative Evaluation of an Internal Flaw of a Structural Component Using Electromagnetic Acoustic Transducers
*Shinichi Maruyama, Toshihiko Sugiura, Akihiro Inoue, and Masatsugu Yoshizawa* .......... 21

Elastic Wave Propagation in Circumferential Direction in Anisotropic Pipes
*S. Towfighi, T. Kundu, and M. Ehsani* .......... 29

Underwater Pipeline Inspection Using Guided Waves
*Won-Bae Na and Tribikram Kundu* .......... 37

**ALLOYS AND ADVANCED MATERIALS**

A Model For Fatigue Crack Initiation From Notches
*A. S. Zaki and H. Ghonem* .......... 49

Atomic Force and Ultrasonic Force Microscopic Surface Characterization of Laser Treated Head Sliders
*Carl Druffner, Edward J. Schumaker, Liming Shen, Shamachary Sathish, Ganesh N. Raikar, and Amarjit S. Brar* .......... 57

**COMPOSITE MATERIALS AND STRUCTURES**

Ultrasonic Assessment of Impact Damages In Fiber-Metal Laminated Panels
*Alexis M. Pierides, Yangxiong Liu, and Benjamin Liaw* .......... 65

**LASER, THERMAL AND OPTICAL NDE METHODS**

Transfer Matrix Formulation With Optical Penetration for Axisymmetric Thermoelastic Wave Propagation in Films
*Chen Li, Jiadao Lin, and Cetin Cetinkaya* .......... 71

IR Thermography Scanning Technique in the Assessment of Airport Pavements
*Antonia Moropoulou, Nicolas P. Avdelidis, and Maria Koui* .......... 79

Trends in NDE Using Thermal Inspection Methods
*Norbert G. Meyendorf, Jochen Hoffmann, Liming Shen, and Henrik Rösner* .......... 83

Development of a Health Monitoring Capability for FRP Using Optical Fiber Sensors
*Christopher Cassino, John Duke,and Jason Borinski* .......... 91

**NOVEL SIGNAL PROCESSING TECHNIQUES**

Blind Image Restoration for Ultrasonic C-Scan Using Constrained 4th Order Cumulants
*Uvais Qidwai and Chi-Hau Chen* .......... 95

Stochastic Properties of Anisotropic Materials
*Michael Peterson and Miao Sun* .......... 105

Radiation Beam Modeling and Intelligent Signal Interpretation for Phased Array Ultasonic Inspection
*Sung-Jin Song, Hyeon Jae Shin, and Jeong-Rock Kwon* .......... 113

**AGING AIRCRAFT: CHALLENGES AND TRENDS**
Inspection of Fastener Holes Using Ultrasonic Phased Arrays
*Pamela G. Herzog, Vincent Lupien, James T. Miller, John J. Selman, and Michael Moles* ... 119
Effect of Corrosion Damage of Fatigue Behavior of AL 2024
*Mohammad Khobaib and Michael S. Donley* ... 127
Use of Computational Intelligence For Corrosion Damage Assessment of Aging Aircraft Panels
*M. J. Palakal, R. M. Pidaparti, S. Rebbapragada, J. Huang, and D. T. Peeler* ... 135
Study of Corrosion Damage Under Protective Coatings
*Mohammad Khobaib, Jochen Hoffmann, Shamachary Sathish, and Michael S. Donley* ... 141

**Author Index** ... 149

Proceedings of
2001 ASME International Mechanical Engineering Congress and Exposition
November 11–16, 2001, New York, NY
NDE-Vol. 21

**IMECE2001/NDE-25800**

# ULTRASONIC INTERFACIAL WAVES IN A FINITELY STRAINED ELASTIC STRUCTURE

**D. A. Sotiropoulos**
*Southern Polytechnic State University, Marietta, GA 30060*
*Tel.: (770) 364 7661, e-mail: sotirop@netscape.net*

## ABSTRACT

Ultrasonic interfacial waves in a finitely strained elastic structure are examined. The structure consists of an incompressible surface layer and a compressible half-space both materials being non-linear elastic and isotropic with otherwise general strain energy densities. On the underlying strain which is in general different in the two materials are superposed ultrasonic interfacial waves. The dispersion equation is obtained in explicit form for pre-strain with common principal axes in the two materials. For a small surface layer, the wave speed is derived in explicit form yielding material parameter conditions for the non-existence of interfacial waves.

## INTRODUCTION

The ultrasonic non-destructive evaluation and characterization of pre-stressed layered materials is important for the structural performance, integrity and failure of the structure. To this end, interfacial ultrasonic waves are a good tool to evaluate interfacial bonds. Of particular interest here are non-linear elastic materials as are, for example, rubber-like materials and biological tissues.

The layered structure considered here consists of an incompressible surface layer overlying a compressible half-space. Both are non-linear elastic materials, finitely strained. Infinitesimal interfacial ultrasonic waves are superposed and their propagation is examined. To make the mathematics tractable, the principal axes of pre-strain are taken to be common in the two materials with one of them perpendicular to the planar interfaces.

Ogden and Sotiropoulos (1995) considered a structure consisting of an incompressible strained surface layer overlying an incompressible strained infinite material and examined the propagation of interfacial waves. Sotiropoulos and Sifniotopoulos (1995) examined interfacial waves propagating between a strained incompressible interlayer and a strained incompressible infinite surrounding material. More recently, Sotiropoulos and Sifniotopoulos (2001) considered the same problem as that in their (1995) paper but with compressible materials.

In the present paper and for the structure and materials considered, the dispersion equation is obtained in explicit form. Its analysis yields the wave speed in explicit form for small surface layers. Material and pre-strain parameter conditions are also obtained for the non-existence of interfacial waves.

## MATHEMATICAL FORMULATION

A surface layer of incompressible material and of arbitrary uniform thickness overlying a compressible half-space is considered. The two materials are non-linear elastic and isotropic and subjected to finite strains. The principal strain directions in the two materials are aligned with one direction being normal to the planar interfaces. The surface layer lies between $x_2 = 0$ and $x_2 = h$, with h denoting its thickness, in a rectangular Cartesian coordinate system ( $x_1, x_2, x_3$), the axes of which coincide with the principal strain directions. The principal stretches are denoted by $\lambda_1$, $\lambda_2$, $\lambda_3$ and $\lambda_1^*$, $\lambda_2^*$, $\lambda_3^*$ in the half-space and surface layer respectively. All quantities referring to the surface layer will be supercripted by a star. The compressibility and incompressibility conditions corresponding for the two materials yields

$$\lambda_1\lambda_2\lambda_3 = J, \qquad \lambda_1^*\lambda_2^*\lambda_3^* = 1 \tag{1}$$

with $J=\rho_r/\rho$ where $\rho_r$ and $\rho$ are the densities of the half-space before and after the finite strain is applied. Ultrasonic interfacial harmonic waves of small amplitude are superposed on the underlying deformation, with motion in the $(x_1, x_2)$ plane and direction of propagation the $x_1$- axis, along the interface. They are represented by their velocity components

$$v_1 = A \exp(skx_2 - ikx_1 + i\omega t)$$

.

$$v_2 = B \exp(skx_2 - ikx_1 + i\omega t) \tag{2}$$

inside the half-space, and by their velocity potential

$$\psi^* = A^* \exp(- s^*kx_2 - ikx_1 + i\omega t) \tag{3}$$

inside the surface layer. $k= \omega/c$ is the wavenumber with $\omega$ being the circular frequency and c the phase speed. The factor s defines the decay of the interfacial wave away from the interface, so that it must have positive real part.

Substitution of one of equations (2) and (3) respectively in the equations of motion of the two materials results in the characteristic equations for $s^2$ and $s^{*2}$:

$$\alpha_{22}\gamma_2\, s^4 - [2\beta-\rho_r c^2\, (\alpha_{22} + \gamma_2\, )]\, s^2 + (\alpha_{11} - \rho_r\, c^2\, )\, (\gamma_1 - \rho_r\, c^2\, ) = 0 \tag{4}$$

$$\gamma^* s^{*4} - (2\beta^* - \rho^*\, c^2\, )\, s^{*2} + \alpha^* - \rho^*\, c^2 = 0 \tag{5}$$

The material constants $\alpha_{11}$, $\alpha_{22}$, $\beta$, $\gamma_1$, $\gamma_2$ and $\alpha^*$, $\beta^*$, $\gamma^*$ are defined in terms of the strain energy density and the principal stresses and stretches of each material. Their definition can be found in Sotiropoulos and Sifniotopoulos (1995), (2001). The decay factors s and $s^*$ are then given as solutions of equations (4) and (5) respectively as

$$s_1^2 + s_2^2 = [2\beta - \rho_r c^2\, (\alpha_{22} + \gamma_2)] / (\alpha_{22}\gamma_2\, ),$$

.

(6)

$$s_1^2 s_2^2 = (\alpha_{11} - \rho_r c^2)(\gamma_1 - \rho_r c^2) / (\alpha_{22}\gamma_2)$$

and

$$s_1^{*2} + s_2^{*2} = (2\beta^* - \rho^* c^2)/\gamma^*, \quad s_1^{*2} s_2^{*2} = (\alpha^* - \rho^* c^2)/\gamma^* \qquad (7)$$

The material constants satisfy the strong ellipticity conditions:

$$\alpha_{11}, \alpha_{22}, \gamma_1, \gamma_2 > 0, \qquad \beta + (\alpha_{11}\alpha_{22}\gamma_1\gamma_2)^{1/2} > 0 \qquad (8)$$

$$\alpha^*, \gamma^* > 0, \qquad \beta^* + (\alpha^*\gamma^*)^{1/2} > 0 \qquad (9)$$

from which it is concluded that $s^2$ and $s^{*2}$ are either real or complex conjugate. Using (2) and (3) and the above, the interfacial waves in the half-space are given as

$$v_1 = [A_1 \exp(s_1 k x_2) + A_2 \exp(s_2 k x_2)] \exp(-ikx_1 + i\omega t)$$

$$v_2 = [B_1 \exp(s_1 k x_2) + B_2 \exp(s_2 k x_2)] \exp(-ikx_1 + i\omega t)$$

(10)

with

$$i s_m \delta B_m = (\rho_r c^2 + \gamma_2 s_m^2 - \alpha_{11}) A_m, \quad m = 1,2 \qquad (11)$$

in which $\delta^2 = \alpha_{11}\alpha_{22} + \gamma_1\gamma_2 - 2\beta$. In the surface layer the interfacial waves are given by

$$\psi^* = [A_1^* \exp(-s_1^* k x_2) + A_2^* \exp(-s_2^* k x_2) + A_3^* \exp(s_1^* k x_2) + A_4^* \exp(s_2^* k x_2)] \cdot \exp(-ikx_1 + i\omega t) \qquad (12)$$

The dependence of the wavespeed, c, on the non-dimensional wavenumber, kh, and on the material and strain parameters will be determined by using the boundary conditions at the interface ($x_2 = 0$) and at the top boundary ($x_2 = h$) of the surface layer. At $x_2 = h$, the traction rate vanishes, i.e.,

$$\gamma^* (\psi^*_{,22} - \psi^*_{,11}) + \sigma_2 \psi^*_{,11} = 0 \qquad (13)$$

$$(2\beta^* + \gamma^* - \sigma_2)\psi^*_{,112} + \gamma^* \psi^*_{,222} - \rho^* \psi^*_{,2tt} = 0 \qquad (14)$$

where comma denotes differentiation and $\sigma_2$ the Cauchy principal stress perpendicular to the layering. At $x_2 = 0$, the velocity and traction rate are continuous, i.e.,

$$v_1 = v_1^*, \; v_2 = v_2^* \quad (15)$$

and

$$(v_{1,2} + v_{2,1}) \gamma_2 / J = (v^*_{1,1} + v^*_{2,1}) \gamma^* \quad (16)$$

$$v_{1,11} \alpha_{12} / J + v_{2,21} \alpha_{22} / J = -v^*_{1,11} (2 \beta^* + \gamma^* - \sigma_2) - v^*_{1,22} \gamma^* + v^*_{1,tt} \rho^* \quad (17)$$

## THE DISPERSION EQUATION

$v_1$, $v_2$ and $\psi^*$ are substituted in the above six boundary conditions resulting in a set of six linear homogeneous algebraic equations for the six constants $A_1$, $A_2$, $A_1^*$, $A_2^*$, $A_3^*$, $A_4^*$. For a nontrivial solution the determinant of their coefficients must vanish yielding the dispersion equation in a determinant form. Its solutions give those wavenumbers for which interfacial waves exist for given material and pre-stress parameters.

The dispersion equation is obtained in explicit form by carrying out the operations in the six by six determinant. The result is:

$$C(\eta, \eta^*, \sigma, r) \sinh^2 [ kh(s_1^* + s_2^*) / 2] / (s_1^* + s_2^*)^2$$

$$- C(\eta, -\eta^*, \sigma, r) \sinh^2 [ kh(s_1^* - s_2^*) / 2] / (s_1^* - s_2^*)^2$$

$$+ D(\eta, \eta^*, \sigma, r) \sinh [ kh(s_1^* + s_2^*)] / (s_1^* + s_2^*)$$

$$- D(\eta, -\eta^*, \sigma, r) \sinh [ kh(s_1^* - s_2^*)] / (s_1^* - s_2^*) + E(\eta, \eta^*, \sigma, r) = 0 \quad (18)$$

where the coefficients C, D and E are defined as

$$C(\eta, \eta^*, \sigma, r) = 2f^* (\eta^*, \sigma^*) [(1 + \eta \alpha_{22}^{-1/2}) f^* (\eta^*, \sigma^*)$$

$$+ 2r (1 - \sigma - \eta \alpha_{22}^{-1/2} \alpha_{12} \gamma_1^{1/2}) (1 - \sigma^* - \eta^*)$$

$$+ r^2 \gamma_1 f(\eta, \sigma) / (1 - \eta^2)] \quad (19)$$

$$D(\eta, \eta^*, \sigma, r) = r (\alpha_{22}^{1/2} \eta + \eta^*) (s_1 + s_2) f^* (\eta^*, \sigma^*) \quad (20)$$

$$E(\eta, \eta^*, \sigma, r) = 2r^2 \eta^* \gamma_1 f(\eta, \sigma) / (1 - \eta^2) \quad (21)$$

with

$$f^*(\eta^*, \sigma^*) = \eta^{*3} + \eta^{*2} + (2\beta^* - \alpha^* + 2 - 2\sigma)\eta^* - (1 - \sigma^*)^2 \quad (22)$$

$$f(\eta, \sigma) = \alpha_{12}^2 \alpha_{22}^{-1/2} \eta^3 + [\alpha_{11} - 1 + (1-\sigma)^2 / \gamma_1] \eta^2 + [\alpha_{22}(\alpha_{11} - 1) - \alpha_{12}^2] \alpha_{22}^{-1/2} \eta - (1 - \sigma)^2 / \gamma_1 \quad (23)$$

$$(s_1^* \pm s_2^*)^2 = \eta^{*2} \pm 2\eta^* + 2\beta^* - \alpha^* \quad (24)$$

$$\eta^2 = (1 - \rho_r c^2) / (\alpha_{11} - \rho_r c^2), \quad \eta^2 = \alpha^* - \rho^* c^2 \quad (25)$$

and

$$\alpha_{11} = \alpha_{11}/\gamma_1, \quad \alpha_{22} = \alpha_{22}/\gamma_2, \quad \gamma_1 = \gamma_1/\gamma_2, \quad \alpha_{12} = \alpha_{12}/(\gamma_1, \gamma_2)^{1/2}, \quad \rho_r = \rho_r/\gamma_1$$

$$\rho^* = \rho^*/\gamma^*, \quad \alpha^* = \alpha^*/\gamma^*, \quad \beta^* = \beta^*/\gamma^*, \quad \sigma = J\sigma_2/\gamma_2, \quad \sigma^* = \sigma_2/\gamma^*, \quad r = \gamma_2/(J\gamma^*) \quad (26)$$

The wave speed, c, is obtained as solution of the dispersion equation (18) for given materials, pre-stress conditions, and wavenumbers.

## SMALL SURFACE LAYER

For kh<<1, a small surface layer, equation (18) to the first power in kh becomes

$$r\gamma_1 f(\eta, \sigma) + (s_1 + s_2)\, g(\eta, \sigma^*, R)\, kh = 0 \quad (27)$$

where

$$g(\eta, \sigma^*, R) = \alpha_{22}^{1/2} [R\alpha_{11} - 2\beta^* - 2(1 - \sigma^*)] \eta^3 + [R\alpha_{11} - \alpha^* + (1 - \sigma^*)^2] \eta^2 + \alpha_{22}^{1/2} [2\beta^* - R + 2(1 - \sigma^*)] \eta + \alpha^* - R - (1 - \sigma^*)^2 \quad (28)$$

and $R = \rho^* / \rho_r$ with $\rho^*$, $\rho_r$ as defined in (26). Now, set in equation (27) $\eta = \eta_0 + kh\eta_1$ where $\eta_0$ is the unique solution of $f(\eta, \sigma) = 0$, the equation governing the propagation of surface waves in a strained compressible homogeneous half-space. Then, the slopes $\eta_1$ is given as

$$\eta_1 = -(s_1 + s_2)|_{\eta = \eta_0} \; g(\eta_0, \sigma^*, R) / [\dot{\gamma}_1 f_\eta(\eta_0, \sigma)] \qquad (29)$$

where the subscript $\eta$ in f denotes differentiation with respect to $\eta$. The phase speed is then given as

$$c = c_0 \{1 - \eta_0 \eta_1 (\alpha_{11} - 1) kh / [r\rho_r c_0^2 (1 - \eta_0^2)^2]\} \qquad (30)$$

in which $c_0$ is the surface wave phase speed that satisfies $f(\eta, \sigma) = 0$ and is given by

$$c_0^2 = c_L^2 (1 - \alpha_{11} \eta_0^2) / (1 - \eta_0^2), \quad c_L^2 = \dot{\rho}_r^{-1} \qquad (31)$$

with $\rho_r$ as defined in (26).

It is concluded from equation (27) that when $\alpha_{11} > 1$ sufficient conditions for the existence of interfacial waves are

$$\alpha^* - R - (1 - \sigma^*)^2 \leq 0 \quad \text{and} \quad g(\alpha_{11}^{-1/2}, \sigma^*, R) \geq 0 \qquad (32)$$

when $\delta^2 \leq \alpha_{22} (\alpha_{11} - 1)$, and only the second of (32) when $\delta^2 > \alpha_{22} (\alpha_{11} - 1)$. Analogous results are derived for $\alpha_{11} < 1$.

It is revealing to consider further the case $\sigma = 1$. If $\delta^2 \leq \alpha_{22} (\alpha_{11} - 1)$ and $\alpha_{11} > 1$ then $\eta = 0$ is a solution of $f(\eta, 1) = 0$ and, therefore, for small kh we have

$$\eta = \eta_1 kh / r \qquad (33)$$

with

$$\eta_1 = \gamma_1^{-1/2} [\alpha_{22} (\alpha_{11} - 1) - \delta^2]^{-1/2} [R - \alpha^* + (1 - r)^2] \qquad (34)$$

By definition, $\eta_1$ must be positive in order for the interfacial wave to decay away from the interface. Therefore, it is concluded from (34) that interfacial waves cannot be propagated under the condition

$$R - \alpha^* + (1 - r)^2 \leq 0 \qquad (35)$$

This implies that

$$R \geq \alpha^* \qquad (36)$$

is a sufficient condition for the existence of propagating interfacial waves between a strained compressible half-space underlying a strained incompressible surface layer.

## REFERENCES

**Ogden, R. W. and Sotiropoulos, D. A.** (1995), "On interfacial waves in pre-stressed layered incompressible elastic solids", *Proc. R. Soc. Lon. A,* Vol. 450, pp. 319-341

**Sotiropoulos, D. A. and Sifniotopoulos, C. G.** (1995), "Interfacial waves in pre-stressed elastic interlayers", *J. Mech. Phys. Solids,* Vol. 43, pp. 365-387

**Sotiropoulos, D. A. and Sifniotopoulos, C. G.** (2001), "The effect of stress on interfacial waves in elastic compressible interlayers", in *Mechanical Waves for Composite Structures Characterization* (edited by D. A. Sotiropoulos), Kluwer Academic Publishers, pp. 168-186

Proceedings of
2001 ASME International Mechanical Engineering Congress and Exposition
November 11–16, 2001, New York, NY
NDE-Vol. 21

IMECE2001/NDE-25801

# USE OF LAMB WAVES TO PREDICT CURRENT DAMAGED STATE OF MORTAR

Michael S. Keller

Tribikram Kundu

Chandrakant S. Desai

Center for Material Modeling and Computational Mechanics, Department of Civil Engineering and Engineering Mechanics, University of Arizona, Tucson, AZ

## ABSTRACT

This paper describes the experimental and analytical results for the development of a methodology for predicting the current state of simulated concrete (mortar) in infrastructure. The nondestructive Lamb wave technique is used to experimentally identify the disturbance (damage) before and after mechanical testing of mortar specimens, which is used in the initial research as a substitute for concrete. The mortar specimens are tested to measure their stress-strain response under uniaxial compression and tension loading. The results of the nondestructive and mechanical stress-strain testing are correlated to develop the model based on the Disturbed State Concept (DSC), a unified approach for modeling material behavior. This model can allow evaluation of the deformation moduli, strength and degradation (damage) at a given state during the life of the material. This information can be used to design rehabilitation strategies. It can also lead to the development of new computer based equipment that can be used in the field for defining the remaining life. At this time, the research involved one-dimensional testing. The proposed methodology can, however, be extended and improved by conducting two- and three-dimensional testing of concrete specimens.

## INTRODUCTION

The deformation and strength properties of structural materials such as concrete, metals, and asphalt in engineering structures modify, (usually deteriorate) due to fatigue, wear and tear under applied loads and environmental effects (temperature, chemical attack, weather, etc.). As a consequence, it may happen that after a certain period of service life, the structure requires evaluation of its current condition, and as necessary, design of rehabilitation based on the deformation moduli and strength left during the service life.

A number of field measurement techniques have been used to make an evaluation of the condition of a structure. Nondestructive techniques [1-7] are among the most commonly used. These methods usually allow for the computation of elastic properties such as Young's Modulus, E, from NDT properties. However, most available techniques do not provide information on the existing state of stress that can allow calculation of reduced values of moduli, and the remaining strength. Since both the deformation and strength properties are needed in the design for rehabilitation, it is desirable to develop a methodology that can permit evaluation of the entire stress-strain response including deformation and strength components corresponding to the current condition.

## REVIEW OF PREVIOUS RESEARCH

In the inspection of metal materials and structures, Non Destructive Testing (NDT) is an accepted practice. Ultrasonic and radiographic techniques have been used for some time to locate damage in the material, and there are recognized national and international standards on their use [5]. On the other hand, NDT of concrete is technologically far behind NDT of steel because of the complications involved with particle size, inhomogeneity, and fabrication. Steel and other metals are formed at plants, which adhere to controlled standards of production, whereas concrete is largely placed and cured in the field using relatively unskilled labor.

One of the most popular and significant applications of NDT has been to find flaws and determine the size and shape of flaws inside of a material. This is typically done for metals (specifically welds in pipes/pressure vessels) to determine if manufacturing or fabrication processes are done satisfactorily. It is proposed here, that this methodology can be extended on a more global scale such that many flaws are detected by a single test over a significant length of the material. In typical flaw detection, the wavelength used may be quite small compared to the size of the flaw. This works well because the wave is reflected by the material-flaw interface (or the surface of the flaw). On a global scale, however, short wavelengths will be scattered by these flaws, and the resulting signal can be noisy and inconsistent. A wavelength, which is large compared to the flaw size, is required to determine the *overall* effect of flaws in the material. In this way, microscopic flaws, such as micro-cracks, or small voids due to particle motion (slip and rotation) can be quantified. For example, flaw density or amount of damage may be determined and assigned a value. Since a material develops flaws (micro-cracks or voids) as it is loaded, and the damage (or disturbance) is a function of the amount of flawed (or broken) material, it follows that the disturbance can be determined from Lamb wave Testing.

One difficulty of NDT and especially of NDT for concrete is the attenuation of the measured wave. Lamb waves and other guided waves do not attenuate as quickly as bulk waves, and therefore, give better and more consistent results over a distance in the material. In addition, it has been shown that bulk waves are not affected greatly by changes in stress-strain properties [8], whereas Lamb wave characteristics can be altered greatly by small changes in these properties.

For these reasons, Lamb wave technique has been chosen as the tool with which to measure the internal damage in the mortar material. By varying the angle and frequency of the incident wave, different lamb modes can be generated; these different modes may be more sensitive to different types of flaws leading to a more complete detection of internal damage.

Whitcomb et. al. [9] performed a series of acoustic tests and concluded that ultrasonic waves with frequency up to 510 kHz will propagate through 10 cm of concrete 0.8 cm aggregate size, while a larger aggregate concrete requires a lower frequency to propagate the same distance. It is generally accepted that low frequency waves are required to propagate through granular materials because high frequency (short wavelength) waves are scattered by the aggregate matrix interfaces. The mortar of the experiment conducted here has maximum aggregate size of about 0.40 cm, and therefore 500 kHz transducers were used. The wavelength of a 500 kHz wave traveling at 2 km/s is 0.40 cm.

## NEW CONTRIBUTIONS

The notion that Lamb wave testing can be used to evaluate damage in concrete is a relatively new concept [10-13] Although nondestructive testing (NDT) has been used for many years, the current research is different in two major aspects. First, the experimental setup excites specific Lamb modes by controlling frequency and striking angle, and second, the research focuses on combining and integrating the NDT results with a powerful constitutive model to predict the current damage state of the material. Previous methods (referenced above) involve evaluation only of elastic (stiffness) moduli and are not capable of providing realistic evaluation of the current condition including degree of failure, plastic deformation and degradation.

The objective of the research has been to develop a new and unique procedure for defining the entire stress-strain (pre and post peak) response during the life of structural materials like concrete based on results of NDT. In this manner, both the deformation and strength properties can be evaluated. The procedure developed herein integrates nondestructive (Lamb wave) techniques and the Disturbed State Concept (DSC) for constitutive modeling of materials; it is called the NDT-DSC method. The DSC is formulated and integrated with NDT for the characterization of the current state of the material; details of the DSC are given later. Measurements such as dispersion curves and attenuation for different frequencies and incidence angles are obtained from the Lamb wave technique. The mechanical properties such as the state of stress and elastic modulus are obtained from uniaxial compression and tension tests on specimens of mortar. The nondestructive and mechanical data is correlated so as to develop the methodology for predicting the current state of the material.

This exploratory research can lead to the development of field equipment in which the Lamb wave measurements are input in the DSC model that is installed on a computerized set-up. Then the design parameters for the current state prediction can be obtained directly from the field equipment.

## NDT-DSC METHOD

The DSC is fundamental and mechanistic, and at the same time, it can be simplified for practical applications, [14]. For instance, the DSC is hierarchical and as a result, various previous models such as elastic, elastoplastic, elastoviscoplastic, and continuum damage are available as special cases, depending on the specific material and user need.

Figure 1 shows a schematic of the stress-strain response of a material that is affected by mechanical forces and environmental effects. The elastic modulus, which represents the slope, $E^i$ of the unloading response is often used to compute deformation. The maximum or peak stress, $\sigma^p$, is a basis for the strength used in design.

A brittle material like mortar or concrete under loading may experience microcracking sometime before the peak is reached. Increasing load may cause further microcracking. The microcracks form, grow, and coalesce to form distinct fractures leading to initiation of failure. At a point in the material in an engineering structure during its service life, the state of stress can be anywhere between the initial and failure states. The objective is to develop the NDT-DSC method that can permit evaluation of the current condition (modulus, stress) of the material.

The DSC method allows for the evaluation of the current or observed state of stress ($\sigma^a$) on the basis of the knowledge of the material's initial (relative intact – RI) and ultimate (fully adjusted – FA) states. As a simplification at this time, the behavior is defined in terms of the initial and ultimate elastic moduli, $E^i$ and $E^c$ respectively. The current response is obtained by using the interpolation and coupling function, called the disturbance function (D). Hence, if the disturbance is defined from NDT and stress-strain response, the current stress, $\sigma^a$ can be obtained, and so also the current modulus, $E^a$ (Fig. 1). Details of the NDT and DSC and the procedure for finding the current state, which can be used in the current state prediction are given below.

## EXPERIMENTAL STUDY

Quasi-static (loading, unloading, reloading) tests were conducted on mortar specimens to determine stress-strain response and elastic properties. Mortar is used in this initial research as a substitute for concrete. Separate samples were tested nondestructively using Lamb waves to find the "intact" properties of the material. These samples were loaded to failure and then tested again nondestructively to determine how the samples were affected by the loading. A correlation between the pre- and post-peak properties is made using the Lamb wave analysis, and the stress-strain data.

### Specimens

All of the samples (dimensions shown in fig. 2) were made from mortar with an average 28-day compressive strength of 20 MPa (2920 psi). The mortar was of type N with a sand cement ratio of 3:1 and a water cement ratio of 0.65. For testing and reporting purposes, this material was labeled material 'A'. A block of mortar was poured and allowed to cure outdoors for 48 hours. Then the block was placed in a humidity room and cured for two weeks. The block was then removed from the curing room so that samples could be cut from it. The samples were cut with a large enclosed diamond blade saw that uses a water-oil mixture for lubricant. Two strain gauges were attached to each sample; one in the axial direction, and one in the lateral direction. The strain gauges measured 30 mm (~1.25 in) long and were glued to the specimens using a fast drying epoxy. The location of the gauge was measured and marked; then the epoxy was smeared around the area and the gauge was then held in place until the epoxy was dry. The epoxy was allowed to cure for 24 hrs before any testing was performed; in no case did the strain gauge de-bond from the sample. The samples were capped with a hard ceramic called hydro-stone.

## Testing

**Mechanical Test Procedure.** All tests were displacement controlled. The test setup is shown schematically in Fig. 3. Each sample was loaded to peak with unloading at various points (displacement intervals) to determine elastic modulus at these points. The loading continued until the samples could no longer support a nominal load. The elastic moduli of the samples decreased with loading.

**Nondestructive Test Procedure.** The Nondestructive testing was performed with a Wavetek 100 MHz Arbitrary Waveform Generator (Model 395) controlled by a PC. Transducers were used in an immersed pitch-catch arrangement. The computer served as both the control and data acquisition devices, Fig. 4. The Wavetek generated signal was amplified before exciting the transducer, and the received signal was filtered before entering the computer (A/D converter).

The NDT samples were subjected to Lamb wave frequency sweep testing. The transducers used had a central frequency of 500 kHz, and a frequency scan was performed for each of five incident angles ($10^o$, $15^o$, $20^o$, $25^o$, $30^o$) in order to plot the Amplitude vs. Frequency (V(f)) curves. The range of the scan was between 100 kHz and 1000 kHz for all samples and all angles. The different angles were used to excite different Lamb modes.

## Dispersion Curves

Figure 5 shows a plot of the theoretical dispersion curves for a 1.27 mm (0.5 in) thick mortar plate free on both surfaces. The derivation of the equations used to develop these curves can be found in Schmerr [15] and Mindlin [16]. Each curve represents a "Lamb mode" which may be excited for the given geometry and material properties. Many different vibration modes, symmetric and asymmetric, may be excited for the specimens under consideration. The angles used correspond to different phase velocities calculated from Snell's Law as described below. For a given angle, the phase velocity is a constant value for any frequency as indicated by the horizontal black lines in Fig. 5. Figure 5a and 5b are for intact and damaged samples respectively. Note that there are many more possible modes for a damaged (FA) sample. Snell's Law is given by

$$\frac{V_i}{\sin(\theta_i)} = \frac{V_t}{\sin(\theta_t)} \tag{1a}$$

where $V_i$ is the velocity of the incident wave, $V_t$ is the velocity of the transmitted wave, $\theta_i$ is the incident angle (measured from a line perpendicular to the surface of the sample), and $\theta_t$ is the angle of the transmitted wave. For Plate waves produced through immersion, $V_i$ is the longitudinal wave speed in water (1.49 km/s), and the angle of the transmitted wave is $90^o$. This simplifies the equation to

$$V_t = \frac{1.49}{\sin(\theta)} \text{ km/s} \tag{1b}$$

Each Sample was subjected to a Lamb wave frequency scan before and after being loaded mechanically such that the voltage amplitude vs. frequency, V(f), curves were obtained for the undamaged (relative intact - RI) state, and for the damaged (fully adjusted - FA) state of the sample.

## Test Results

From the two compression tests, the average values of the intact Elastic modulus ($E^i$) and Poisson's ratio ($\nu^i$) were found to be 8.62 GPa (1250 ksi) and 0.19, respectively. The average value of the elastic modulus (for the two compression samples) near the ultimate condition ($E^c$) was found to be 0.724 GPa (105 ksi). The Poisson's Ratio was assumed to be 0.49 due to the small volume change that is associated with plastic deformation; it could not be measured near the ultimate condition because the strain gauges were broken. The theoretical dispersion curves in Fig. 5 were generated based on these elastic properties. Figure 6 shows the Lamb wave V(f) scans for different incidence angles for the sample A-NDT1. The measured amplitude vs. frequency results relate to the intact specimen before testing and the "failed" or damaged sample at the end of testing.

## Discussion of Test Results

Figure 7 shows three significant measurements from the wave testing of specimen A-NDT1; the number of peaks, area under V(f) curve, and maximum peak amplitude. A number of observations can be made based on Fig. 7. The number of peaks detected after failure is much higher than the number of peaks detected before loading. Referring to Fig. 5, it is apparent that a damaged (FA) sample can propagate many more Lamb modes than an intact (RI) specimen; this implies that there will be a greater number of peaks on the V(f) curve for a damaged specimen. The increase in the number of peaks is, therefore, a strong indication that the material has experienced degradation or damage.

There is a decrease in the area under the V(f) curve, Λ, in the failed material. Although the area under the V(f) curve is not in units of energy, it can be related to the transmitted energy. It is clear that for each incident angle, Λ is significantly smaller after loading. This is an indication of the amount of internal damage the sample has experienced.

The maximum amplitude of the scan decreases after loading. For each incident angle, the maximum peak before loading is much higher than the maximum peak after loading. For most of the angles, the post failure V(f) curve had to be magnified in order to identify the peaks.

The post loading V(f) scan for angles less then 25 degrees gave values of peak amplitude and Λ which were hardly distinguishable from the noise in the system. Although this shows that the sample is broken, the results are not reliable, and it is recommended that only results from $25^o$ and $30^o$ be used. Note also that the larger incident angles correspond to slower traveling waves, Eqs. 1a and 1b. A smoothing function has been applied to the graphs in order to establish trends in the data. This approximation serves well to measure $\Lambda$ as well as showing one more aspect of the NDT; namely the peaks are shifted to the right or left after the sample has experienced some damage.

## MATERIAL MODEL

Numerous constitutive models exist today for the characterization of the three-dimensional stress-strain behavior of materials. Many of these models are complicated and relate only to specific features of material behavior such as elastic, plastic or damage. The Disturbed State Concept (DSC) requires the definition of two material responses called reference states, and only one parameter called disturbance. It is a unified, yet simplified approach for the characterization of significant features, elastic, plastic, and creep strains, microcracking, damage, and softening, in a single framework [14]. Comparisons of the DSC with other models, such as damage, fracture, and micromechanical, together with its advantages in terms of generality and simplification for practical use are given by Desai [14].

The DSC is based on the idea that a deforming material can be considered to be composed of Relative Intact (RI) or continuous and Fully Adjusted (FA) parts. The FA parts are due to cracking, slippage, rotations, etc. [14]. Because the material is discontinuous the conventional definition of Stress = Force/ Area does not apply. Instead, a weighted value of stress defined as Force / A` where A` is the nonlocal area that includes the effects of discontinuities in the area where stress is evaluated. The DSC defines the exhibited behavior in terms of the responses of the RI and FA parts through a coupling and interpolation function called a disturbance function.

The RI state can be defined in a number of ways. It can be defined, for example, as the linear elastic response neglecting microcracks or it may be defined as the elastoplastic behavior neglecting friction which would cause loss of energy. A constitutive model such as the $\delta_0$ version in the hierarchical single surface (HISS) concept is often used to represent the RI state [14], which includes irreversible or plastic deformations. The relative intact response is relative in the sense that it excludes the effect of factors which would cause the material to deviate from its initial intact state. Therefore, the choice of the constitutive model to define the RI state will depend on the type of material, and laboratory data which will allow characterization of the necessary material parameters.

As a simplification, the linear elastic response can be used to represent the RI response. In this case, the elastic modulus, E, and the Poisson's ratio, $\nu$, will be all that is required to define this behavior. In this study, the RI response is simulated by using the elasticity model.

As the material is loaded, (i.e. loaded, unloaded, etc.) parts of the material may change from the RI state to the FA state. The location and amount of the FA portions of the material will depend on the initial state of the material, residual stresses, flaws, discontinuities, and microcracking. At the local level, the material particles may experience instantaneous instability leading to movement (both translation and rotation) Fig.8. The measured response at the macro level will integrate all of the material responses at the micro level to give an overall behavior, which can be modeled using the DSC. The FA material can be characterized in a number of ways. If it is considered to be a constrained liquid-solid (constrained by the RI material), then it may be able to resist both shear and normal stresses. If the FA material is considered to be like a constrained liquid, then it will only be able to resist normal stresses, and will have no shear strength. Finally, if it is considered to be a crack or void, then it will not be able to resist shear or normal stresses, and it has no strength at all, which is consistent with the assumption in the classical damage model [17]. Note that this allowance for an FA state by the DSC model is very important, and is what sets this model apart from classical damage models. The assumption that damaged material carries no stress at all may be valid at or near failure. The effect of the damaged portions of a material on its stress-strain behavior is generally significant. The disturbance function allows for this effect and also for the effect of interaction between the FA and the RI portions.

For the general three-dimensional case, constitutive equations with disturbance are

$$\sigma_{ij}^{a} = (1-D)\sigma_{ij}^{i} + D\sigma_{ij}^{c} \tag{2}$$

where $\sigma_{ij}$ is the stress tensor, the superscripts i, a, and c correspond to the RI, observed, and FA behaviors respectively, and D is the disturbance assumed to be a scalar in a weighted sense. The incremental form of Eq. (2) is given by

$$d\sigma_{ij}^{a} = (1-D)\sigma_{ij}^{i} + D\sigma_{ij}^{c} + dD(\sigma_{ij}^{c} - \sigma_{ij}^{i}) \tag{3}$$

or

$$d\sigma_{ij}^{a} = (1-D)C_{ijkl}^{i}d\varepsilon_{kl}^{i} + DC_{ijkl}^{c}d\varepsilon_{kl}^{c} + dD(\sigma_{ij}^{c} - \sigma_{ij}^{i}) \tag{4}$$

where $C_{ijkl}$ is the constitutive tensor and dD is the increment or rate of disturbance. The one-dimensional specialization of Eq. (2) is given as

$$\sigma^{a} = (1-D)\sigma^{i} + D\sigma^{c} \tag{5}$$

which can also be written as

$$\sigma^{a} = (1-D)E^{i}\varepsilon^{i} + DE^{c}\varepsilon^{c} \tag{6}$$

where E is the elastic modulus, and $\varepsilon$ is the strain. If it is assumed that the strains in the RI and FA parts are equal, i.e. $\varepsilon^{i} = \varepsilon^{c}$, then

$$\sigma^{a} = [(1-D)E^{i} + DE^{c}]\varepsilon \tag{7}$$

Then the observed elastic, E, and shear, G, moduli can be expressed in terms of the disturbance as

$$E^{a} = (1-D)E^{i} + DE^{c} \tag{8}$$
$$G^{a} = (1-D)G^{i} + DG^{c} \tag{9}$$

If it is assumed that $E^{c} = G^{c} = 0$, i.e. the damaged portion has no strength,

$$E^a = (1-D)E^i \tag{10}$$
$$G^a = (1-D)G^i \tag{11}$$

Alternatively, the disturbance may be expressed in terms of stress or elastic modulus as

$$D_\sigma = \frac{\sigma^i - \sigma^a}{\sigma^i - \sigma^c} \tag{12a}$$
$$D_E = \frac{E^i - E^a}{E^i - E^c} \tag{12b}$$

where $E^a$ denotes observed modulus at a given state, Fig 1, and the subscripts E and σ denote disturbance based on stress and elastic modulus, respectively. If $E^c = \sigma^c = 0$, then

$$D_E = \frac{\sigma^i - \sigma^a}{\sigma^i} \tag{13a}$$
$$D_E = \frac{E^i - E^a}{E^i} \tag{13b}$$

### CORRELATION BETWEEN DSC AND NONDESTRUCTIVE BEHAVIOR FROM LAMB WAVE MEASUREMENTS

A relation between the disturbance parameter and some aspect of non destructive measurement is desired such that the Lamb wave results can be correlated to the DSC to predict the current state of a material. It stands to reason that the area under the V(f) curve or Λ, for example, can be used to define the disturbance, and therefore, the location of the current state of the material on the stress-strain response. Once this location is known, the current state can be approximated. At this time, V(f) curves, and therefore values of Λ, are known for a specimen before and after testing (i.e. $\Lambda^i$ and $\Lambda^c$). Then the disturbance parameter can be defined as

$$D = \frac{\Lambda i - \Lambda^a}{\Lambda^i - \Lambda^c} \tag{14}$$

where Λ denotes area under V(f) curve.

The disturbance may be defined using Eq. (12b), where E is defined from the NDT rather than the mechanical testing. The modulus, E, can be theoretically defined from the shear and compression wave velocities according to one-dimensional wave propagation theories [15] or from comparison of the measured dispersion curves to theoretical curves generated using a specific E value. Referring to Figure 7, and assuming that Λ after failure corresponds to a disturbance value near unity, a linear relationship between elastic modulus and Λ can be derived as

$$E = \frac{\Lambda^a - \Lambda^c}{\Lambda^i - \Lambda^c}\left(E^i - E^c\right) + E^c \tag{15}$$

Although the disturbance can be calculated directly from Λ, there is some advantage to this more general procedure in that the elastic modulus, E, may be determined in other ways as described above.

### Initial and FA Moduli and Stress

For the prediction of material behavior, the initial elastic modulus, $E^i$, can be calculated from available tests at the time of initial design or it may be found from specimens cast and tested for the composition of the material (concrete) used in the structure under investigation. The values of $E^c$ and $\sigma^c$ can be based on laboratory testing or they can be assumed; they can often be assumed zero for simplicity. The values of $\Lambda^i$ and $\Lambda^c$ must initially be found from Lamb wave testing of the material in question before and after mechanical testing. Once these properties are found, however, they can be considered material properties. Then a correlation can be made between the disturbance calculated from NDT properties, $D_\Lambda$, and the disturbance calculated from mechanical loading $D_\sigma$. Using this correlation, it is possible to predict the location of the current state of the material on the stress-strain response from Eq. (7).

Following is an example showing the step-by-step procedure for determining: 1) the stress-strain response, 2) the location of the current state of the material on the stress-strain response, and 3) the peak stress and strain. This analysis uses the stress-strain data from the mortar samples.

### Example I Prediction of Current Material State on Stress-Strain Response:

**Part I – Prediction of Stress-Strain Response.** It has been shown that Lamb waves can be used to detect damage within a thin grout plate. A material model is now desired to complete the process such that nondestructively measured parameters can be associated with the current state of damage in the material and the current state may be estimated. The following procedure uses foregoing figures and derivations to show that the elastic modulus may be used to determine the disturbance parameter, D, which can then be used to predict the current state.

The stress-strain response of sample A-Comp2 is shown in Fig. 9. Data points taken from this graph are used to develop material parameters. The initial (RI) and final (FA) Elastic moduli are calculated from the stress-strain graphs and are given as

| | |
|---|---|
| $E^i = 1250000$ psi | Initial Elastic Modulus |
| $E^c = 105000$ psi | Final Elastic Modulus |

and the FA and RI stress-strain responses are, respectively, given by

$$\sigma_k^i = E^i \cdot \varepsilon_k^{\ a} \tag{16}$$
$$\sigma_k^c = 60 \tag{17}$$

where k is a counter denoting increment – not to be confused with d in Eq. (3). The values are evaluated at a number of points, and k is the number of each point. The elastic, $\varepsilon_e$, and plastic, $\varepsilon_p$ strains at any point are given by

$$\varepsilon_{e_k} = \frac{\sigma_k^a}{E^i} \tag{18}$$

$$\varepsilon_{p_k} = \varepsilon_k^a - \varepsilon_{e_k} \tag{19}$$

where $\varepsilon_k$ is the total strain at point k. The disturbance is then calculated according to Eq. (12a)

$$D_{\sigma_k} = \frac{\sigma_k^i - \sigma_k^a}{\sigma_k^i - \sigma_k^c}$$

In the functional form, D is often defined in terms of the plastic strain, $\varepsilon_p$, by using a form of Weibull function as

$$D_\sigma = D_u(1 - e^{-A\varepsilon_p^Z}) \tag{20}$$

where A and Z are material parameters to be defined, and $D_u$ is the ultimate disturbance (usually 1.0). Details of the derivation and procedure for determining A, Z, and $D_u$ are given by Desai [14]. To summarize, a plot of ln(-ln(1-D/Du) vs. ln($\varepsilon_p$) must be generated and a straight line is fitted to the data points, Fig 10. The slope of this line leads to Z, and

$$A = e^Y \tag{21}$$

where Y is the y-intercept of the line. A plot of disturbance vs. plastic strain is shown in Fig. 11. Eq. (5) can then be used to back predict the stress-strain response where the disturbance is calculated from Eq. (20). Figure 12 shows that the back prediction compares very well with the test data.

**Part II – Prediction of Current Location on Stress-Strain Response.** To validate the model, an attempt is made to predict a *separate and independent* stress-strain response using the disturbance equation and the RI response above.

Figure 13 shows the stress-strain response for sample A-Comp1, which is made from the same batch of grout as A-Comp2, and the resulting prediction. The prediction procedure is similar to part I, but this time, it is assumed that $\sigma_{c_k} = 0$ such that the only properties required for the prediction are the initial elastic modulus, and A and Z.

In order to evaluate plastic strains, an iterative scheme is required. Substitution of Eqs. (16), and (20) into Eq. (5) gives

$$\sigma^a = [1 - D_u(1 - e^{-A\varepsilon_p^Z})]E^i\varepsilon^a + D_u(1 - e^{-A\varepsilon_p^Z})\sigma_c \tag{22}$$

Substituting Eq. (22) into Eq. (18) gives

$$\varepsilon_e = \frac{1}{E^i}\{[1 - D_u(1 - e^{-A\varepsilon_p^Z})]E^i\varepsilon^a + D_u(1 - e^{-A\varepsilon_p^Z})\sigma_c\} \tag{23}$$

and substituting Eq. (23) into Eq. (19) gives the iterative scheme required.

$$\varepsilon_p = \varepsilon^a - \frac{1}{E^i}\{[1 - D_u(1 - e^{-A\varepsilon_p^Z})]E^i\varepsilon^a + D_u(1 - e^{-A\varepsilon_p^Z})\sigma_c\} \tag{24}$$

Here, the parameters E, A, and Z determined before are used. The value of $\varepsilon_p$ which satisfies Eq. (24) can now be found by iteration. The elastic strain may then be calculated by rearranging Eq. (19) as

$$\varepsilon_e = \varepsilon^a - \varepsilon_p \tag{25}$$

and the stress-strain response can be predicted using the disturbance function from Eq. (5). Of primary importance is the fact that this graph has been derived using only the disturbance parameters defined for this material and the assumed RI and FA behaviors. The total strain increment is arbitrary. In addition, now that the disturbance is known for any point, the plastic strain is known, and also an expression for the elastic modulus is given by Eq. (8)

$$E^a = (1 - D)E^i + DE^c$$

For the mortar material tested, the average initial and final moduli were given before and a relation for disturbance in terms of E can now be calculated according to Eq. (12b)

$$D_E = \frac{E^i - E^a}{E^i - E^c}$$

Thus for any value of E, the disturbance, and therefore, plastic strain can be found by rearranging Eq. (20) to give

$$\varepsilon_p = \left[\frac{-1}{A}\ln\left(1 - \frac{D}{D_u}\right)\right]^{\frac{1}{Z}} \tag{26}$$

The total strain is then found by substitution of Eqs. (16) and (18) into Eq. (19), and solving for $\varepsilon^a$

$$\varepsilon^a = \frac{\varepsilon_p}{D} - \frac{\sigma^c}{E^i} \tag{27}$$

The stress is then calculated from the disturbance Eq. (5), where the disturbance is now expressed as a function of E as in (14b).

$$\sigma^a = (1 - D_E)E^i\varepsilon + D_E\sigma^c \tag{28}$$

Then for a given value E, the current condition of the material for the predicted stress-strain response can be identified. As stated previously, the value of E may be determined from the area under the V(f) curve or $\Lambda$. For verification, we will use the data for sample A-NDT1 where the initial and final values of $\Lambda$ are:

$\Lambda^i = 3265$, $\Lambda^c = 1170$, and let $\Lambda^a = 1180$

Then, from Eq. (17)

$$E = \frac{\Lambda^a - \Lambda^c}{\Lambda^i - \Lambda^c}\left(E^i - E^c\right) + E^c$$
$$E^a = 110465$$

The predicted stress and strain are then

$\varepsilon^a = 0.033$
$\sigma^a = 197$

And the location on the stress-strain response is shown in Fig. 14. It can, therefore, be concluded that, for a given material with known properties, if the current elastic modulus can be determined (through nondestructive testing), then the current state of the material can also be determined using the method described above. Field validation of the elastic modulus is possible using a hammer impact test or similar method.

**Part III – Prediction of Peak Stress and Strain.** The peak stress and peak strain may be found from conventional methods. Assuming $D_u = 1$, and rearranging Eq. (22) gives

$$\sigma^a = (e^{-A\varepsilon_p^Z})(E^i\varepsilon_p + \sigma^a) + (1 - e^{-A\varepsilon_p^Z})\sigma^c \tag{29}$$

solving for $\sigma^a$ gives

$$\sigma^a = \frac{E^i\varepsilon_p + e^{A\varepsilon_p^Z}\sigma^c - \sigma^c}{e^{A\varepsilon_p^Z} - 1} \tag{30}$$

differentiation with respect to $\varepsilon_p$

$$\frac{d\sigma^a}{d\varepsilon_p} = -E^i\,\frac{(-\varepsilon_p e^{A\varepsilon_p^Z} + \varepsilon_p + AZe^{A\varepsilon_p^Z}\varepsilon_p^{(Z+1)})}{\varepsilon_p(e^{A\varepsilon_p^Z} - 1)^2} \tag{31}$$

then, setting this equation to zero, the value of the plastic strain at the peak is determined from

$$-\varepsilon_p e^{A\varepsilon_p^Z} + \varepsilon_p + AZe^{A\varepsilon_p^Z}\varepsilon_p^{(Z+1)} = 0 \tag{32}$$

Equation (32) can be evaluated using a root solver; note that this equation is for the specialized case when the RI response is assumed to be linear elastic. Once the peak plastic strain is found, the disturbance at peak or $D_p$ can be evaluated from Eq. (20); $D_p$ can then be used to find peak total strain, $\varepsilon_{peak}$, from Eq. (27) and peak stress, $\sigma_p$, from Eq. (5).

## DESIGN BASED ON NDT

Using the values determined in Part III of Example I, the design value for peak stress can be determined by the following relation (see Fig. 1):

If $D > D_p$ Design peak stress = $\sigma^a$
If $D < D_p$ Design peak stress = $\sigma_p$

Elastic modulus and Shear modulus can be determined from Eqs. (8) and (9).

## FIELD EQUIPMENT

The exploratory research outlined here can lead to the development of field equipment to examine materials in a structure, predict its stress-strain response, and specify design parameters for rehabilitation - all in the field. A schematic representation of this equipment is presented in Fig. 15. An outline of the procedure, which would be carried out by this equipment, is given below.

**Procedure to predict the current state of concrete structures using Lamb wave techniques and the Disturbed State Concept.**

1. Define RI and FA behavior
2. Determine the material parameters A and Z – these may be assumed or calculated based on what *is* known of the material.
3. Formulate disturbance function in terms of plastic strain using A and Z → Eq. (20)
4. Back predict stress-strain response (optional).
5. Pick strain values and find corresponding plastic strain from Eq. (24), corresponding disturbance from Eq. (20), and corresponding stress from Eq. (5). Use these values to predict the Stress-Strain response.
6. Determine E or Area under V(f) curve, $\Lambda$, for structure under consideration, using NDT methods. Determine disturbance value for structure under consideration using

Eqs. (12b) or (14).

7. Calculate position on predicted stress- strain response from Eqs. (5) and (27).
8. Calculate peak stress according to Example I part III
9. Calculate design peak stress according to:

   If $D > D_p$ Design peak stress = $\sigma_a$

   If $D < D_p$ Design peak stress = $\sigma_p$
10. Calculate E and G from Eq. (8) and (9).

## SUMMARY AND CONCLUSIONS

From the analysis presented herein, the following conclusions can be made

1. A Lamb wave frequency scan is capable of detecting damage inside a mortar plate.
2. The information produced by the scan can be used in a number of ways to define the amount of internal damage or disturbance of the material.
3. The Disturbed State Concept (DSC) can be used in combination with stress-strain data to predict the stress-strain behavior of other samples of the same material.
4. The disturbance parameter determined from Lamb wave testing can be used in combination with the DSC to identify the location of the current state of the material on its predicted stress-strain response.
5. The current state of the sample may be approximated based on the previous determinations.

Although in its exploratory stages, the combination of nondestructive testing of concrete type materials, and the DSC has shown to be a powerful tool. Further research and development are required before the concept can be finalized. When it is, however, the development of equipment, which may be able to determine the remaining life of bridges and other infrastructure can be developed.

## ACKNOWLEDGEMENTS

A part of the research in this paper was supported by the IDEA project No. NCHRP[1]-54.

## REFERENCES

1. Jones, R., 1962, "Surface wave technique for measuring the elastic properties and thickness of roads: Theoretical development," British Journal of Applied Physics, **13**, pp. 21-29.
2. Heisley, J. S., Stokoe, K. H. III, and Meyer, A.H. 1983, "Moduli of pavement systems from analysis of surface waves," Transp. Res. Record, 852, TRB, Washington, DC, 22-31.
3. Ioannides, A. M. 1990, "Dimensional analysis in NDT rigid pavement evaluation," J. of Transportation Engineering, ASCE, 116(1), 23-36.
4. Krause, M., Wiggenhauser, H., Barmann, O., Langenberg, K., Frielinghaus, R., Wollbold, F., Schichert, M., 1995, "Comparison of Pulse-Echo Methods for Testing Concrete", Proceedings of the International Symposium on Nondestructive Testing in Civil Engineering (NDT-CE), **1**, Berlin, Germany, DGZFP, pp. 281-296.
5. Malhotra, V. M., and Carino, N. J., 1991, "CRC Handbook on nondestructive Testing of Concrete", CRC Press.
6. Pessiki, S. and Johnson, M., 1994, "Nondestructive Determination of Concrete Strength in Plate Structures by the Impact Echo Method", Review of Progress in Quantitative Nondestructive Evaluation, **13**, pp. 2139-2146.
7. Sansalone, M. J., and Strett, W. B., 1997, "Impact-echo: Nondestructive Evaluation of Concrete and Masonry", Bullbrier Press.
8. Nagy, P. B., Adler, L., Mih, D., and Sheppard, W., 1989, "Single mode Lamb wave inspection of composite laminates," Review of Progress in Quantitative NDE, D.O. Thompson and D. E. Chimenti (eds), plenum Press, New York, **8B**, pp. 1535-1542.
9. R. W. Whitcomb, L. Jacobs, and L. Aref 1992, "Quantitative Ultrasonic Evaluation of Concrete" Proceedings of the International Conference on Nondestructive Testing of Concrete in the infrastructure, pp 238-255.
10. Jung, Y. C., Na, W. B., Kundu, T. and Ehsani, M. R., 2000, "Damage Detection in Concrete Using Lamb Waves", Nondestructive Evaluation of Highways, Utilities, and Pipelines IV, Ed. A. E. Aktan and S. R. Gosselin, Proc. Of SPIE NDE 2000, March 5-9, 2000, Newport Beach, California, **3995**,pp.448-458.
11. Kundu, T., Ehsani, M., Maslov, K. I. and Guo, D., 1999, "C-Scan and L-Scan Generated Images of the Concrete/GFRP Composite Interface", NDT&E International, **32**, pp.61-69.
12. Na, W. B., Kundu, T. and Ehsani, M. R., 2000, "Inspection of Concrete-Metal Rod Interface Using Guided Waves", IMECE 2000, Orlando, Florida, Nov.5-10.
13. Popovics, J. S., 1994, "Some Theoretical and Experimental Aspects of the Use of Guided Waves for the Nondestructive Evaluation of Concrete", Ph.D. Dissertation, The Pennsylvania State University, University Park, PA.
14. Desai, C. S. 2001, "Mechanics of Materials and Interfaces: The Disturbed State Concept," under publication, CRC Press, Boca Raton, Florida.
15. Schmerr, L. W. 1998, "Fundamentals of Ultrasonic Nondestructive Evaluation," Plenum Press, New York and London.
16. Mindlin, R. D., 1960, "Waves and Vibrations in Isotropic Elastic Plates", in Structural Mechanics (J. N. Goodier and N. J. Hoff, eds.), Pergamon, NY.
17. Kachanov, L. M., 1986, "Introduction to Continuum Damage Mechanics", Martinus Nijhuft Publishers, Dordrecht, The Netherlands.
18. Schickert, G. and Wiggenhauser, H., 1995, "Comparison of DIN/ISO 8047 (Entwurf) to Several Standards on Determination of Ultrasonic Pulse Velocity in Concrete" International Symposium on Non-Destructive Testing in Civil Engineering (NDT-CE) Sept. 26-28, 1995, Berlin. Germany Proceedings **1** Lectures **2**, p. 1346.
19. Jung, Y. C., Kundu, T. and Ehsani, M., 2001, "Internal Discontinuity Detection in Concrete by Lamb Waves", Materials Evaluation, in press.

[1] NCHRP, National Academy of Sciences, Washington, D.C.

20. Na, W. B., Kundu, T. and Ehsani, M., 2001, "Lamb Waves for Detecting Delamination between Steel Bars and Concrete", Journal of Computer-Aided Civil and Infrastructure Engineering, in review.
21. Desai, C. S., and Toth, J. 1996, "Disturbed state constitutive modeling based on stress-strain and nondestructive behavior," Int. J. solids and Structures, 33, 11, 1619-1650.

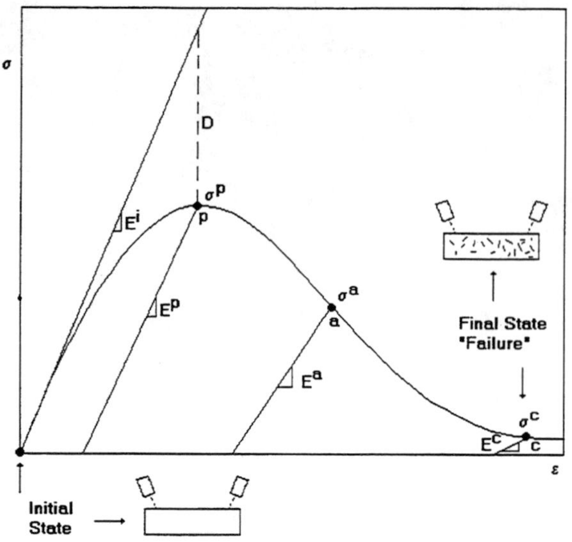


**Figure 1. Schematic of stress-strain response. NDT-DSC concept.**

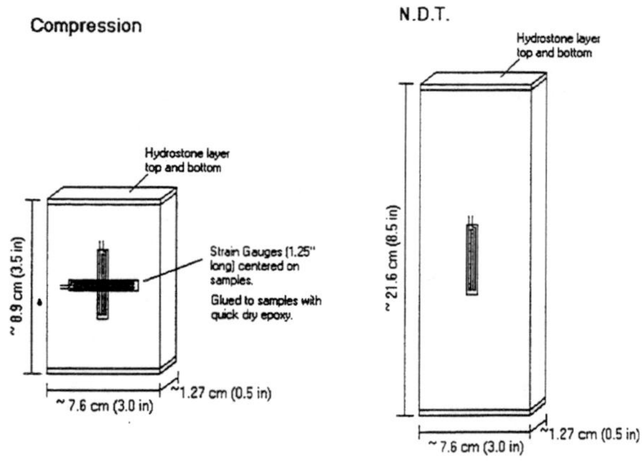


**Figure 2. Sample Dimensions and preparation.**

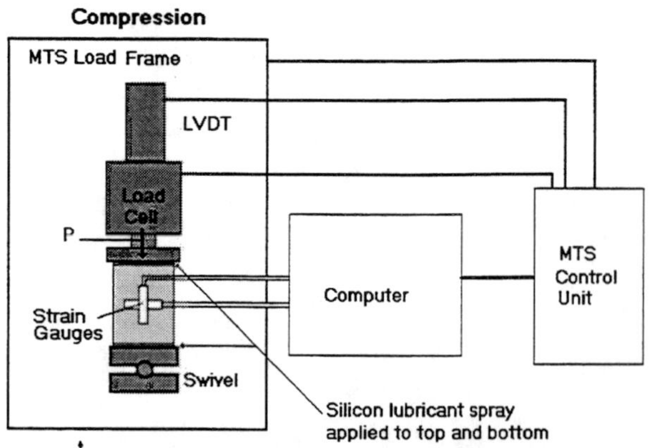


**Figure 3. Mechanical test setup.**

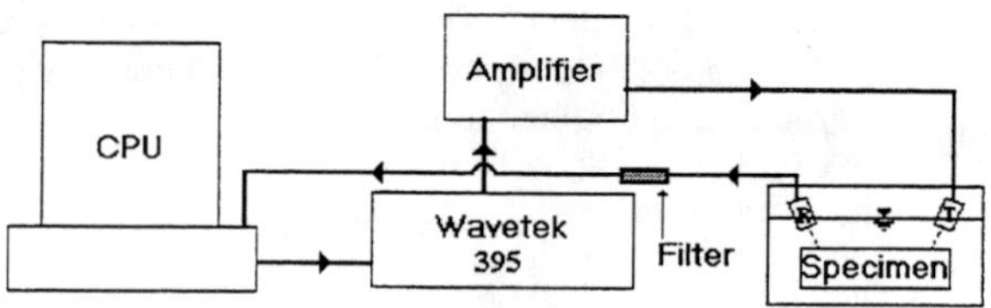


**Figure 4. NDT Setup.**

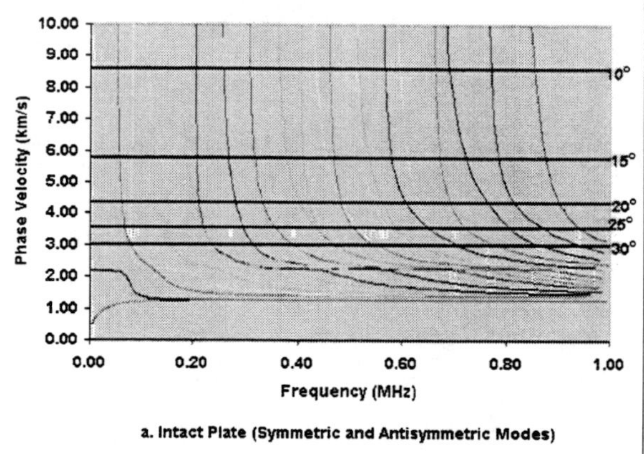


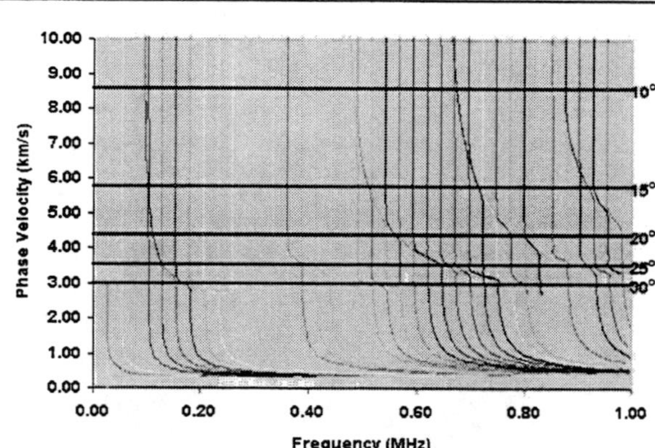


**Figure 5. Theoretical Dispersion curves for 12.7 mm (0.5 in thick) mortar plate.**

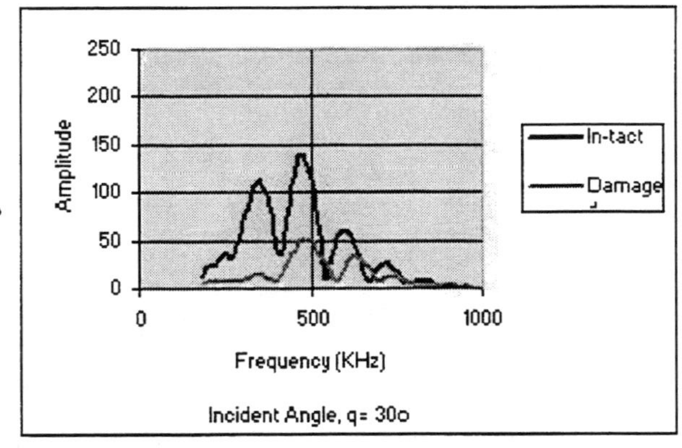


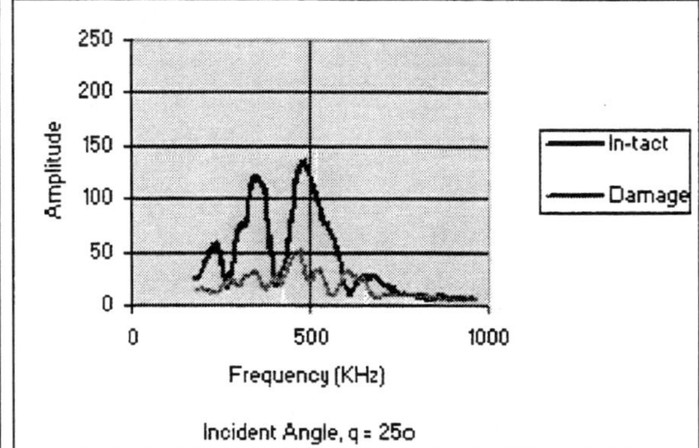


**Figure 6. V(f) curves for sample A-NDT1**

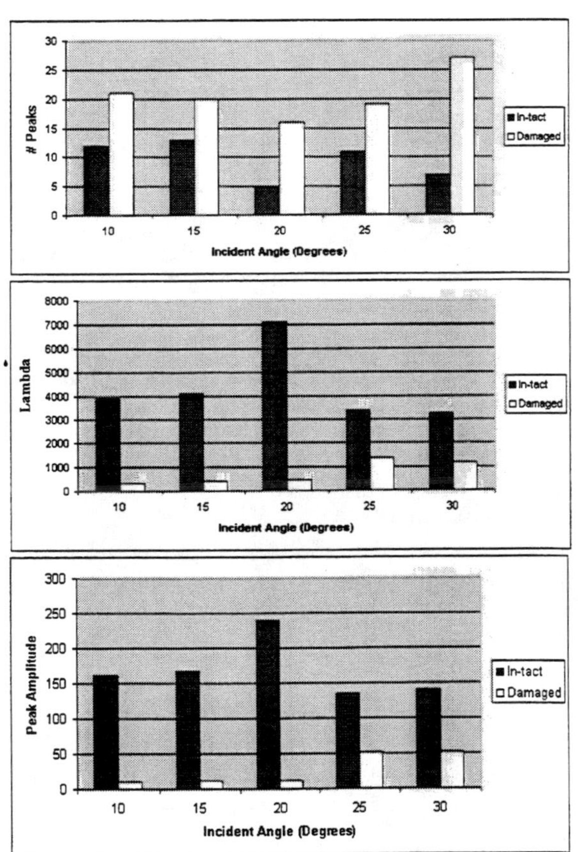


**Figure 7. NDT results for sample A-NDT1**

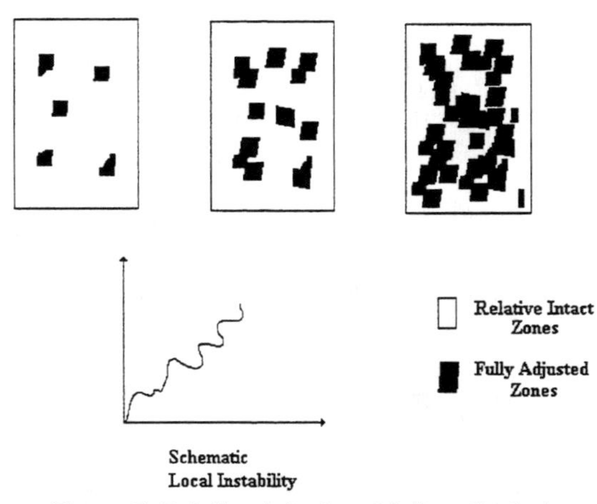


**Figure 8. Relative intact and fully adjusted clusters.**

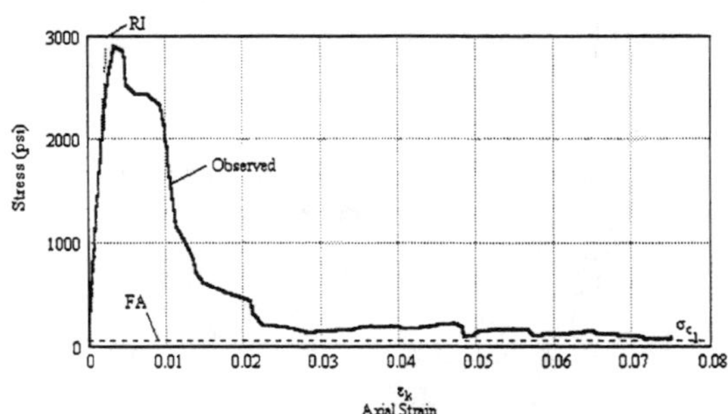


**Figure 9. Stress-strain response of sample A-comp2.**

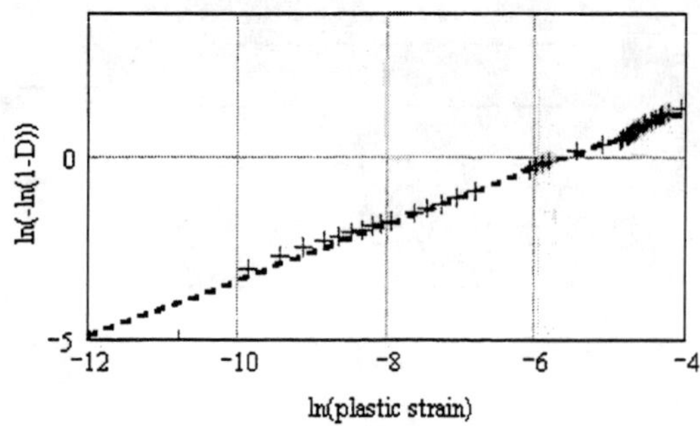

Figure 10. Plot to determine parameters A and Z.

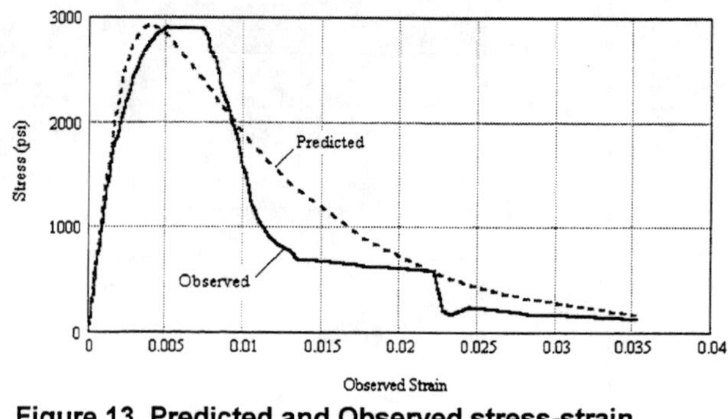

Figure 13. Predicted and Observed stress-strain response of sample A-Comp1.

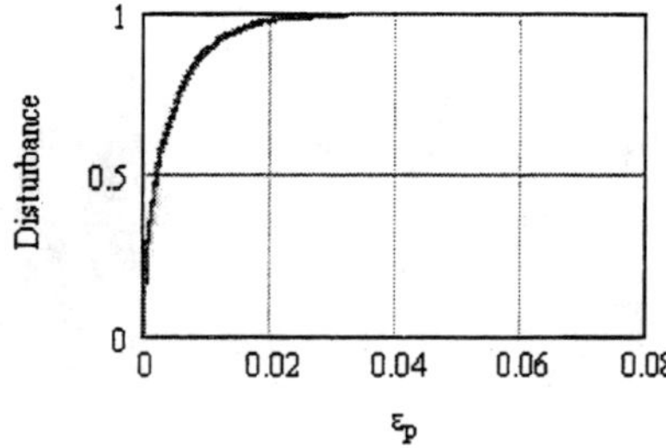

Figure 11. Disturbance vs. plastic strain.

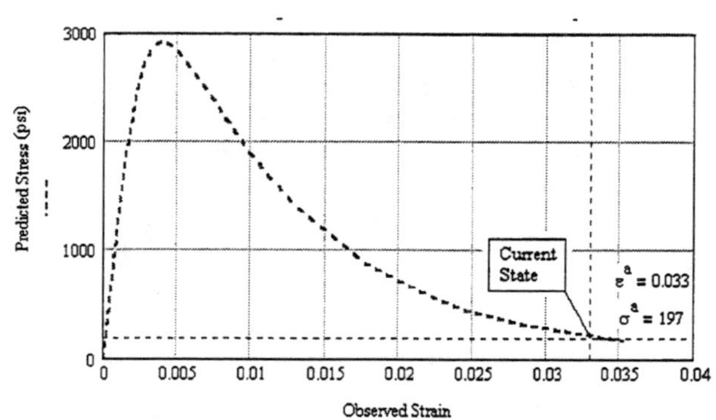

Figure 14. Current state on stress-strain response.

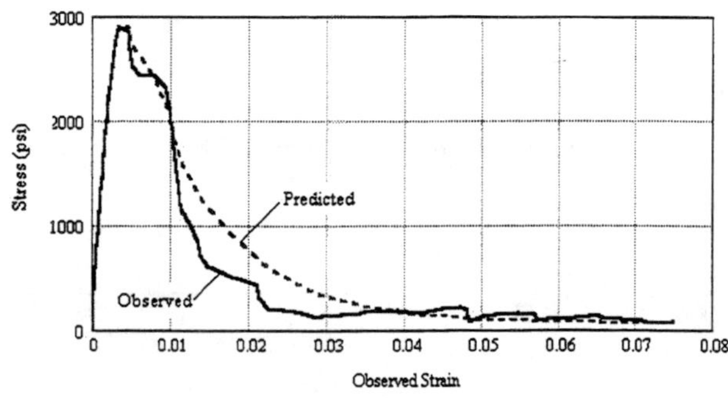

Figure 12. Predicted response of sample A-comp2.

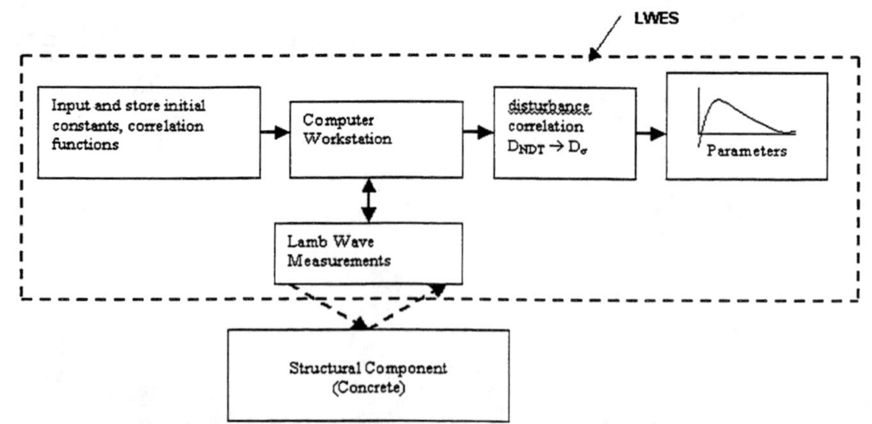

Figure 15. Schematic of Lamb Wave Equipment System (LWES)

Proceedings of
2001 ASME International Mechanical Engineering Congress and Exposition
November 11–16, 2001, New York, NY
NDE-Vol. 21

IMECE2001/NDE-25802

# QUANTITATIVE EVALUATION OF AN INTERNAL FLAW OF A STRUCTURAL COMPONENT USING ELECTROMAGNETIC ACOUSTIC TRANSDUCERS

**Shinichi Maruyama**
Department of Mechanical Engineering
Faculty of Science and Technology
Keio University
3-14-1, Hiyoshi, Kouhoku-ku,
Yokohama, Kanagawa 223-8522, Japan
Email: d996212@msr.st.keio.ac.jp

**Toshihiko Sugiura**
**Akihiro Inoue** and **Masatsugu Yoshizawa**
Department of Mechanical Engineering
Faculty of Science and Technology
Keio University
3-14-1, Hiyoshi, Kouhoku-ku,
Yokohama, Kanagawa 223-8522, Japan
Email: sugiura@mech.keio.ac.jp

## ABSTRACT

Electromagnetic acoustic transducers (EMATs) can transmit and detect ultrasonic waves in a conductive specimen out of any contact with it. This process can be given theoretical modeling and formulation based on elastodynamics and electromagnetics. It suggests a possibility of quantitative nondestructive evaluation using EMATs. This research deals with a numerical method of flaw identification from a receiver signal obtained by EMATs. Experimental results of the receiver signals agree well with numerical ones, which verified the mathematical model of the inspection process. Flaw identification is formulated as a problem of parameter optimization. To avoid being trapped in a local optimum, initial parameters were successfully evaluated from the height and the time period of peaks in the receiver signals. Flaw parameters were identified from the receiver signals obtained by numerical simulations and experiments, which verified the method of flaw identification presented here.

## INTRODUCTION

Quantification and automation of non-destructive evaluation (NDE) techniques are strongly desired to fulfill severe requirements of safety and reliability of structures. Recently, there has been increasing interest in identification of unknown parameters of a flaw by means of inverse analysis of detected signals based on mathematical modeling of the inspection process. So far, there have been various studies on inverse problems in processes of ultrasonic testing (UT), such as identification of a flaw in inverse scattering methods(Hirose, 1993) and application of boundary integral equation method to crack determination problems for time domain elastodynamics(Nishimura, 1994). However, the authors have found no report of inverse analysis of detected signals obtained in UT considering interaction between electric signals of input / output and vibration of an elastic specimen. That is because mathematical modeling of the inspection process is hard to obtain owing to a complex mechanism of transmission of vibration between transducers and the specimen.

On the other hand, electromagnetic acoustic transducers (EMATs) can transmit and detect ultrasonic waves in a conductive specimen out of any contact with it. This process can be given theoretical modeling and formulation based on elastodynamics and electromagnetics(Ludwig, 1994). Experiment and two-dimensional numerical analysis of the inspection process were already done by the authors, which explained mechanism of transmission and detection of ultrasonic waves by EMATs(Sugiura, 2000) and effects of a flaw on the propagation of ultrasonic waves and on receiver signals. Thus, testing by EMATs is a potential method of quantitative nondestructive evaluation of an internal flaw of a structural component based on an approach of inverse analysis.

This research deals with a numerical method of flaw identification from a receiver signal obtained by EMATs. First, mathematical modeling of an inspection system using EMATs in which a specimen has a cylindrical cavity

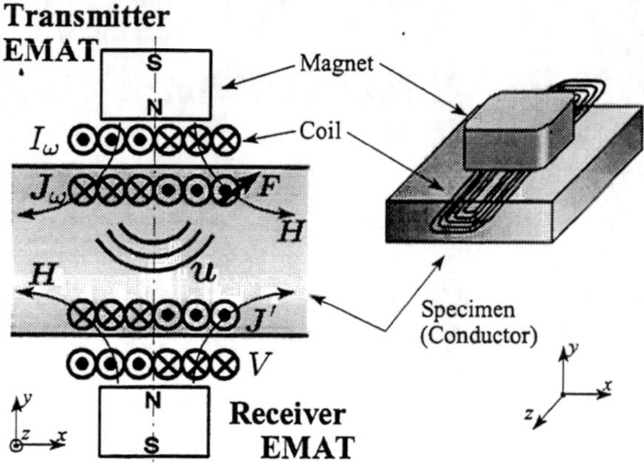


Figure 1. PRINCIPLE OF TESTING BY EMATS

modeling a flaw is described. Computed receiver signals are compared with experimental ones for verification of the mathematical model. Flaw identification is then formulated as a problem of parameter optimization. Finally, validity of the above method is discussed on the basis of results of flaw identification from receiver signals obtained by numerical simulations and experiments.

## MATHEMATICAL MODELING OF AN INSPECTION PROCESS AND VERIFICATION OF IT

### Mechanism of EMAT

Figure 1 shows the principle of testing by EMATs. An EMAT usually consists of a coil for generating time-varying magnetic field and permanent magnets for generating static one. By the time-varying field generated by a.c. current supplied to the coil, eddy current is induced in a boundary layer near the surface of an elastic and conductive specimen. Interaction between the eddy current and the time-varying and static magnetic field causes time-varying electromagnetic force near the surface. This body force generates elastic waves, which propagate in the specimen.

Detecting waves by an EMAT is also possible by the inverse procedure. When the elastic waves reach the receiver side surface, vibration of the conductive specimen under the applied magnetic field generates the electromotive force. This electromotive force makes the eddy current, which induces time-varing magnetic field. As a result, the coil of receiver EMAT detects voltage as a receiver signal.

### Mathematical Modeling of an Inspection System

**An Inspection System** Figure 2 shows an inspection system using EMATs in this research; (a) is the configuration of the system and (b) is a cross section of EMATs and a specimen. By the transmitter system consists of a function generator and a power amplifier one-pulse sinusoidal current at 1MHz with its amplitude 11App is supplied to a coil of the transmitter EMAT. Ultrasonic waves, generated at a surface of the conductive specimen, propagate toward the receiver side, which induce voltage of a coil of the receiver EMAT. The receiver system consists of a preamplifier, a Bessel type low-pass filter of its cut off frequency 1MHz and a digital oscilloscope.

The EMATs used in this research consist of magnets with their residual density of magnetic flux 1.3T and a 7-turn coil. Material parameters of the aluminum specimen used here are the mass density $\rho = 2.69 \times 10^3 \text{kg/m}^3$, the electrical conductivity $\sigma = 4.0 \times 10^7$ S/m, the elastic constants $G = 27.1$ GPa and $\lambda = 55.3$ GPa. Sound speeds of P and S waves in the specimen are $c_d = 6380$m/s, $c_s = 3170$m/s respectively.

The EMATs and the specimen are long in the $z$ direction. Since magnetic field along the $x$ direction is strong near the coils, these EMATs strongly transmit and detect ultrasonic waves those vibrate in the $y$ direction. A cylindrical cavity of diameter $d$ is located in the specimen. The EMATs and the cavity are on a centerline of the specimen ($x = 0$), thus this model is symmetric. Let $h$ be the distance between the center of the cavity and that of the specimen. We assume the cavity is located in $h > 0$. The specimen is sufficiently long also in the $x$ direction for the receiver signals not to be contaminated by the reflected ultrasonic wave from this direction.

**Governing Equations** Since the EMATs and the specimen are sufficiently long in the $z$ direction, we assume electromagnetics and elastodynamics to be two-dimensional. Let $\Omega_{cd}, \Omega_{air}$ and $\Omega_{fl}$ be a region in the conductor, around the conductor and inside the cavity respectively. $\Gamma_{cd}$ and $\Gamma_{fl}$ are surfaces of the conductor and of the cavity. By using symmetric condition, the region to be analyzed reduces to one-half of the specimen, i.e. $x > 0$. The governing equations of those phenomena are given as follows in our two-dimensional model:

$$\frac{1}{\mu_0}\nabla^2 A_{ed} - \sigma\frac{\partial A_{ed}}{\partial t} = \sigma\left[\frac{\partial A_{ex}}{\partial t} - \left(\frac{\partial \boldsymbol{u}}{\partial t} \times \boldsymbol{B}\right)_z\right] \quad (\text{ in } \Omega_{cd}) \quad \text{(1-a)}$$

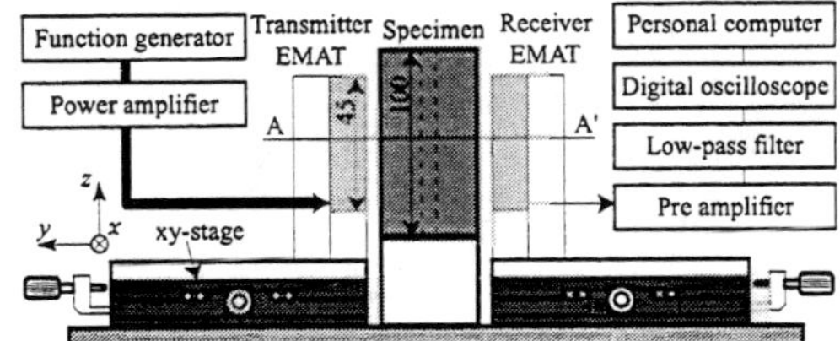

(a) Configuration of inspection system

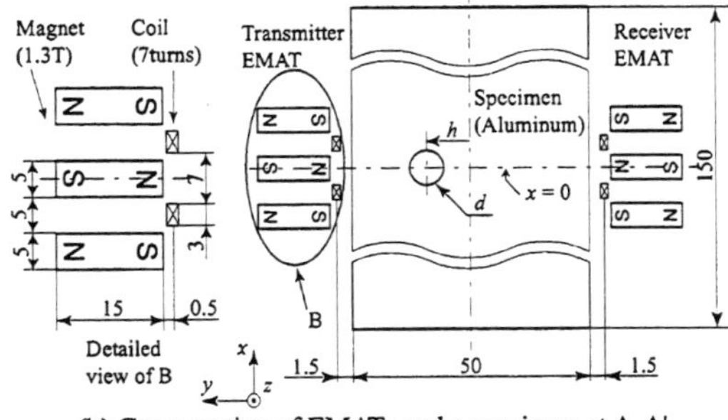

(b) Cross section of EMATs and a specimen at A-A' (2 dimensional numerical model)

Figure 2. INSPECTION SYSTEM USING EMATS

$$\frac{1}{\mu_0}\nabla^2 A_{ed} = 0 \qquad (\text{ in } \Omega_{air}, \Omega_{fl}) \quad (1\text{-b})$$

$$\rho\frac{\partial^2 \boldsymbol{u}}{\partial t^2} - (2G+\lambda)\nabla(\nabla\cdot\boldsymbol{u}) + G\nabla\times(\nabla\times\boldsymbol{u}) = \boldsymbol{J}\times\boldsymbol{B} \qquad (\text{ in } \Omega_{cd}) \quad (2)$$

where $\mu_0$ is the magnetic permeability. Equations (1) are the governing equations of quasi-static electromagnetic field derived from Maxwell's equations. Unknown of these equations is $A_{ed}$ ( $z$ component of magnetic vector potential due to the eddy current in the specimen ). The eddy current $\boldsymbol{J}$ and the magnetic flux density $\boldsymbol{B}$ are expressed as follows:

$$\boldsymbol{J} = -\sigma\left[\frac{\partial A_{ed}}{\partial t} + \frac{\partial A_{ex}}{\partial t} - \left(\frac{\partial \boldsymbol{u}}{\partial t}\times\boldsymbol{B}\right)_z\right]\boldsymbol{e}_z \quad (3)$$

$$\boldsymbol{B} = \nabla\times(A_{ed}+A_{ex})\boldsymbol{e}_z \quad (4)$$

where $A_{ex}$ is the $z$ component of the magnetic vector potential due to the magnet and the coil current. Equation (2) is the displacement equation of motion with a term of electromagnetic external force.

Boundary conditions are as follows: $A_{ed} = 0$ at the infinity and on $x = 0$; traction is free on $\Gamma_{cd}$ and $\Gamma_{fl}$; $u_x = 0, \partial u_y/\partial x = 0$ on $x = 0 \cap \Omega_{cd}$.

#### Numeical Method

For the region in the specimen $\Omega_{cd}$ and $\Omega_{fl}$, Eq. (1) and (2) are converted into a discrete form by the Galerkin finite element method. The boundary element method is used for the electromagnetic field in the air region $\Omega_{air}$. The Newmark $\beta$ method is adopted for time history analysis.

The term $\dot{\boldsymbol{u}}\times\boldsymbol{B}$ is negligible compared to the term $\dot{A}_{ed}$. Thus, electromagnetic field is calculated with the former term neglected at first, and then elastic vibration is solved with the result of electromagnetic force. Using the above results, electromagnetic field induced by the term $\dot{\boldsymbol{u}}\times\boldsymbol{B}$ is solved for the calculation of the voltage of receiver coil. The voltage of the receiver coil can be computed by using the variation of magnetic flux $\Phi$ penetrating the receiver coil. $\Phi$ can be obtained by using $A_{ed}, A_{ex}$ at conductors of the coil.

For the regions $\Omega_{cd}$ and $\Omega_{fl}$, bilinear quadrilateral finite elements are used. $\Gamma_{cd}$ is divided into linear line boundary elements. Total number of nodes is 143,780. $\Delta t = 1/32\mu s$ is chosen for the time step of the time history analysis.

#### Results of Numerical Simulations

Computed receiver signals are shown in Fig. 3 with solid black lines. They are the time histories of the receiver coil voltage, digitally filtered with the same characteristic as that of the low-pass filter used in the inspection system. Figure 3(a) and (e) are the results of $d \sim 4$mm with $h \sim 0$ and 10mm respectively, while (b),(c) and (d) are those of $h \sim 5$mm with $d \sim 4, 6$ and 8mm respectively. A big peak appears directly after $t = 0$, owing to an effect of the electromagnetic field made by the transmitter current. After the big peak, many peaks appears corresponding to arrivals of ultrasonic waves at the receiver side. The diameter and the location of the cavity, $d$ and $h$, make distinct change of these peaks.

#### Verification of Computation by Experiments

Experiments were carried out by using the apparatus shown in Fig. 2. The process of transmission and detection was repeated 2560 times, and the measured signals were averaged. The time history of receiver coil voltage was obtained by considering gain of the pre-amplifier and delay time of the filter. Experimental results of the receiver signals are shown in Fig. 3 by gray lines with their amplitude magnified 1.34 times. They show good agreement with numerical results of the receiver signals. The amplitude of numerical results is 1.34 times bigger than the experimental ones because of difference in distribution of electromagnetic field and in propagation of ultrasonic waves between two-dimensional simulation and the experiment. Thus, numerical and experimental results of the receiver signals show

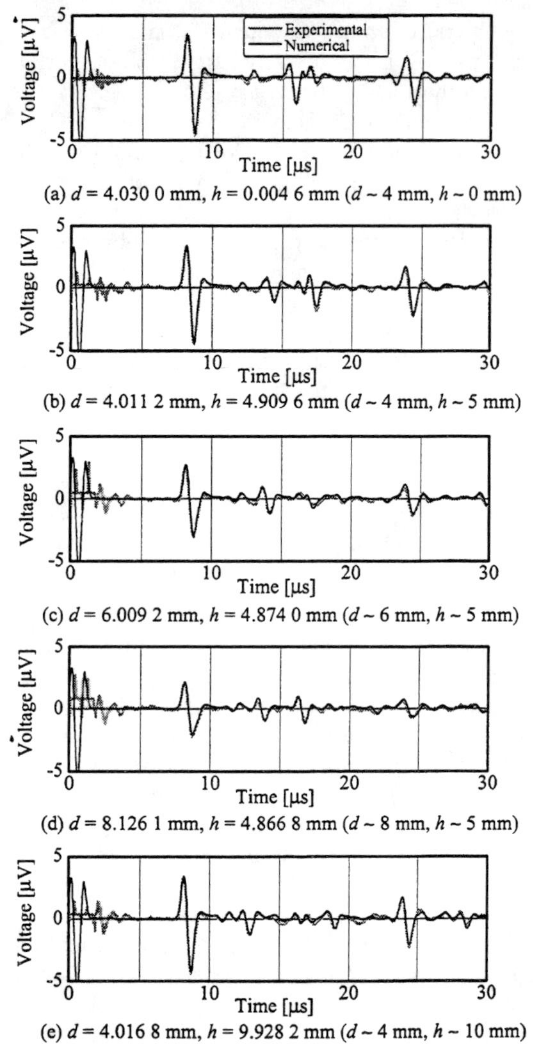


Figure 3. TIME HISTORIES OF RECEIVER COIL VOLTAGE (NUMERICAL AND EXPERIMENTAL RESULTS)

good agreement, which verifies the mathematical model of inspection system and the numerical simulation in this research.

## METHOD OF IDENTIFYING FLAW PARAMETERS

### Formulation of a Problem of Flaw Identification

We assume a problem of flaw identification where a receiver signal $v_m(t)$ can be measured while the diameter and the location of a cylindrical cavity $d, h$ are unknown in the inspection system shown in Fig. 2. The flaw identification can be formulated as a problem of parameter optimization: find a parameter vector of the cavity $\boldsymbol{p} = (d, h)^T$ where a cost function $R(\boldsymbol{p})$ takes a minimum value. $R(\boldsymbol{p})$ is defined as follows:

$$R(\boldsymbol{p}) = \int_{T_1}^{T_2} [v_m(t) - v_c(\boldsymbol{p}; t)]^2 dt \tag{5}$$

where $v_c(\boldsymbol{p}; t)$ is a computed receiver signal assuming that parameters of the cavity are given by $\boldsymbol{p}$. For the identification, we use wave form of the receiver signal corresponding to the arrivals of ultrasonic waves until that of the P wave which reflected at the receiver and the transmitter sides. The big signal due to the transmission is not utilized. Thus, integration in Eq. (5) is calculated from $T_1 = 5\mu s$ to $T_2 = 30\mu s$.

### Optimization

The parameter vector $\boldsymbol{p}$ where the cost function $R(\boldsymbol{p})$ takes a minimum value was calculated by using the conjugate gradient method(Press, 1992). Starting from a given initial value $\boldsymbol{p}_0$, a procedure shown below was iterated.

$$\boldsymbol{p}_{k+1} = \boldsymbol{p}_k + \alpha_k \boldsymbol{s}_k \quad k = 0, 1, 2, \cdots \tag{6}$$

$\boldsymbol{s}_k$ is a vector denoting the search direction computed by using the conjugate gradient method. The coefficient $\alpha_k$, which minimizes $R(\boldsymbol{p}_k + \alpha_k \boldsymbol{s}_k)$, can be calculated by using an arbitrary method of one-dimensional minimization. Brent's method(Press, 1992) is adopted in this research. The gradient of $R(\boldsymbol{p})$, $\nabla R = (\partial R/\partial d, \partial R/\partial h)^T$, is evaluated by calculating values of $\partial v_c/\partial d, \partial v_c/\partial h$ by the finite difference approximation. For example, $\partial R/\partial d$ is obtained as follows;

$$\frac{\partial v_c}{\partial d}(\boldsymbol{p}; t) = \frac{v_c(\boldsymbol{p}'; t) - v_c(\boldsymbol{p}; t)}{\epsilon} \tag{7}$$

$$\left.\frac{\partial R}{\partial d}\right|_{\boldsymbol{p}} = 2 \int_{T_1}^{T_2} [v_c(\boldsymbol{p}; t) - v_m(t)] \frac{\partial v_c}{\partial d}(\boldsymbol{p}; t) dt \tag{8}$$

where $\boldsymbol{p}' = (d + \epsilon, h)^T$ and $\epsilon$ is a tiny quantity.

### Requirement for Initial Guess

The optimization method mentioned above does not always give a global minimum; optimizing parameters converge to a local minimum if the initial values are close to it. A bird's-eye view with its corresponding contour line map in Fig. 4 shows the relation between the cost function and the parameters of the cavity. This cost function is calculated for the computed $v_m(t)$ of $d = h = 5$mm.

Two local minima can be found in the cost function except the global minimum given by the true parameters of

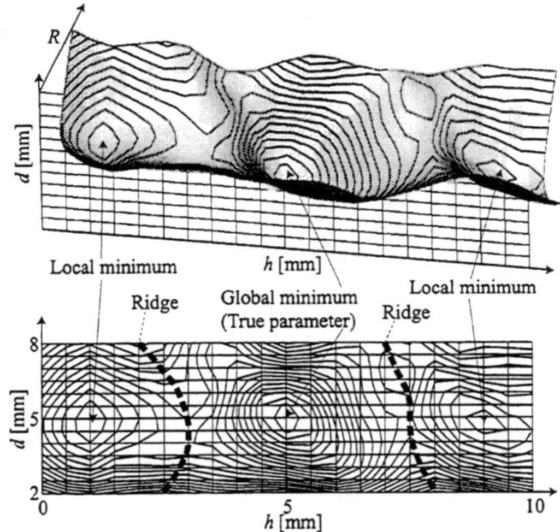


Figure 4. RELATION BETWEEN THE COST FUNCTION R AND THE PARAMETERS OF THE CAVITY

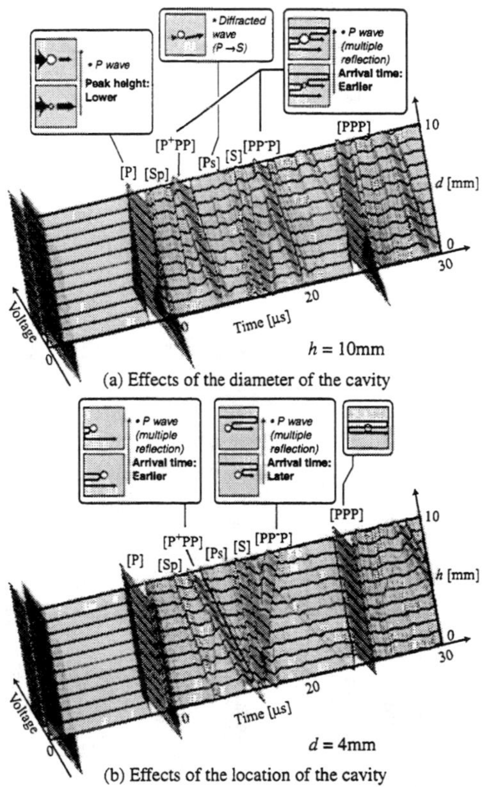


(a) Effects of the diameter of the cavity

(b) Effects of the location of the cavity

Figure 5. RELATION BETWEEN THE RECEIVER SIGNALS AND THE PARAMETERS OF THE CAVITY

the cavity. There are "ridges" those divide the local and the global minima about $\pm$2mm apart from the true parameters in the $h$ direction.

By considering the relation between the waveform of the receiver signal and the parameters of the cavity, existence of these "ridges" can be explained as follows. Figure 5 shows the relation between the receiver signals and the parameters of the cavity. Figure 5(a) represents the dependence of the receiver signal on the diameter $d$ for $h$ = 10mm. Similarly (b) shows that on the location $h$ for $d$ = 4mm. The receiver signals have three kinds of noticeable peaks as follows:

i) [P] and [S] denote peaks due to the arrivals of the P and S waves, respectively, directly propagated from the transmitter side, and [PPP] denotes a peak corresponding to the P wave reflected at the receiver and transmitter sides.
ii) [P$^+$PP] and [PP$^-$P] are due to the arrivals of the P waves after the multiple reflection between the surface of the specimen and that of the cavity.
iii) [Sp] and [Ps] correspond to the detection of diffracted waves whose mode is converted at the surface of the cavity.

Effects of the parameters $d$ and $h$ on these peaks are as follows:

i) [P][S][PPP]: Their peak height is smaller for the bigger $d$ because of the interruption of ultrasonic waves by the cavity. Those peaks are seldom influenced by $h$.
ii) [P$^+$PP][PP$^-$P]: Increase of $d$ shortens the propagation length, thus the peaks appear earlier. Similarly, as $h$ becomes bigger [P$^+$PP] appears earlier, while [PP$^-$P] appears later, corresponding to the path length of the waves.
iii) [Sp][Ps]: As $d$ increases, the height and the width of the peaks slightly increase. For the bigger $h$, [Sp] appears earlier and [Ps] appears later.

Variation in the time of the peaks, [P$^+$PP] and [PP$^-$P], by $h$ influences $R(\boldsymbol{p})$ remarkably. Figure 6 shows the relation between the phase delay $\eta$ of two same shaped peaks and their residual $R$. $R$ takes its maximum value when the phase delay is half of period of the peak, i.e. $\eta = T/2$. Appearance times of [P$^+$PP] and [PP$^-$P] get $2\Delta h/c_d$ earlier or later by the change of $h$, $\Delta h$. The "ridge" of the cost function is generated at $h$ where [P$^+$PP][PP$^-$P] of $v_m(t)$ appear half of the period earlier or later than those of $v_c(t)$.

Therefore, good optimization, avoiding being trapped in a local optimum, requires finding initial values that do not cause phase difference of the peaks due to multiple re-

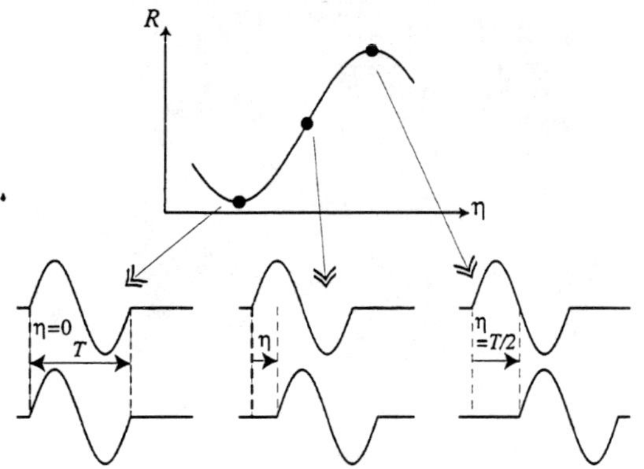


Figure 6. EFFECT OF THE PHASE DELAY OF TWO PEAKS ON THE COST FUNCTION

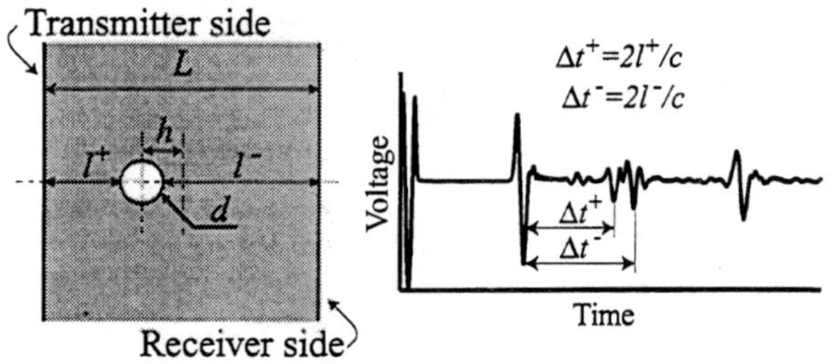


Figure 7. PRINCIPLE OF AN IDENTIFICATION METHOD USING THE EFFECT OF THE MULTIPLE REFLECTION

flection longer than half of the period.

#### Initial Guess by Rough Estimation Using Features of the Receiver Signals

Utilizing time intervals of the peaks due to the multiple reflection of the waves, the parameters of the cavity were roughly estimated to obtain the initial values satisfying the condition mentioned above for the optimization.

The principle of the method described here is shown in Fig. 7. Time intervals of the peaks due to the multiple reflection of the waves, $\Delta t^+$ and $\Delta t^-$, in the receiver signal $v(t)$ can be evaluated by using autocorrelation function $g(\tau)$ as shown below.

$$g(\tau) = \frac{\int_0^T v(t)v(t-\tau)dt}{\int_0^T v^2(t)dt} \tag{9}$$

Since $g(\tau)$ takes its maximum at $\tau = \delta t$ when peaks in $v(t)$ appears with an interval $\delta t$, one can obtain $\Delta t^+$ and $\Delta t^-$. $\Delta t^+$ and $\Delta t^-$ are equal to the time that it takes for the P wave to travel twice the distance from the transmitter side to the surface of the cavity and that from the receiver side to the surface of the cavity, $2l^+/c_d$ and $2l^-/c_d$, respectively. Thus, one can obtain two parameters of the cavity given by $d = L - l^- - l^+$ and $h = (l^- - l^+)/2$, where $L$ stands for the length of the specimen.

The autocorrelation function $g(\tau)$ has many maxima including the ones not corresponding to $\Delta t^+$, $\Delta t^-$. Thus, information of the height of [P] was used to determine $\Delta t^+$ and $\Delta t^-$. First, an approximate value of $d$ was evaluated using the height of [P]. After subtracting waveforms of [S][PPP] corresponding to the approximate $d$ from $v(t)$, $g(\tau)$ was obtained. Averaging the receiver signals of different $h$ cancels the peaks whose time is transformed by $h$, consequently the waveforms of [S][PPP] can be calculated. Second a pair of maxima of $g(\tau)$ was selected as $\Delta t^+$ and $\Delta t^-$ not to contradict the approximate value of $d$.

## RESULT OF FLAW IDENTIFICATION

#### Identification Using Numerical Results of the Receiver Signals

Using computed receiver signals instead of measured signals $v_m(t)$, simulation of the flaw identification was carried out.

**Rough Estimation Using Features of the Receiver Signals** Results of the rough estimation of the cavity are shown in Fig 8. Plots and Solid lines represent estimated parameters and true parameters respectively. They show good agreement. The largest error of the estimation occurs at the true parameters of, $d = 7$mm and $h = 1$mm, because two peaks corresponding to the multiple reflection ([P$^+$PP][PP$^-$P]) are so close that only one maximum appears in $g(\tau)$, i.e. estimated $h$ becomes 0.

Results of the rough estimation obtained here fulfill the condition that the initial values must satisfy in the parameter optimization, since error of the estimation of $h$ is within 1mm.

**Identification by the Method of Optimization** Identification of the cavity by using the method of parameter optimization, using the results of rough estimation mentioned above as initial parameters. A history of optimization is shown in Fig 9. The parameters converge to their true values as the cost function decreases.

Table 1 summarizes the results of rough estimation $h_i, d_i$ and those of optimization $h_e, d_e$ along with indices

Table 1. TRUE AND ESTIMATED VALUES (IDENTIFICATION USING NUMERICAL RESULTS OF THE RECEIVER SIGNALS)

| True values | | Rough estimate | | Index of error | Optimized values | | Index of error |
|---|---|---|---|---|---|---|---|
| $h$ (mm) | $d$ (mm) | $h_i$ (mm) | $d_i$ (mm) | $\delta_i$ (%) | $h_e$ (mm) | $d_e$ (mm) | $\delta_e$ (%) |
| 0.000 0 | 4.000 0 | 0.000 0 | 3.922 4 | 1.940 8 | 0.001 0 | 3.999 8 | 0.050 3 |
| 2.500 0 | 4.000 0 | 2.399 5 | 3.987 7 | 5.036 2 | 2.502 3 | 4.002 9 | 0.136 1 |
| 5.000 0 | 4.000 0 | 4.949 2 | 3.845 5 | 4.623 1 | 4.999 8 | 4.000 9 | 0.025 5 |
| 7.500 0 | 4.000 0 | 7.508 0 | 3.800 6 | 5.000 3 | 7.500 1 | 4.000 1 | 0.007 1 |
| 10.000 | 4.000 0 | 9.904 9 | 3.609 4 | 10.862 | 9.999 5 | 3.997 6 | 0.065 7 |
| 5.000 0 | 5.000 0 | 4.902 9 | 4.740 8 | 6.478 6 | 4.999 5 | 4.999 7 | 0.020 2 |
| 5.000 0 | 6.000 0 | 4.886 6 | 5.627 0 | 7.277 0 | 4.999 7 | 5.998 9 | 0.020 8 |
| 5.000 0 | 7.000 0 | 4.789 0 | 6.624 7 | 8.068 3 | 5.003 2 | 7.010 3 | 0.173 7 |
| 5.000 0 | 8.000 0 | 4.762 8 | 7.539 3 | 8.266 1 | 5.001 2 | 8.003 4 | 0.051 1 |
| 1.000 0 | 7.000 0 | 0.000 0 | 8.235 6 | 33.584 | 0.999 4 | 7.000 7 | 0.020 5 |

$$\delta_i = \frac{\sqrt{\Delta d_i^2 + 4\Delta h_i^2}}{|d|} = \frac{\sqrt{(d_i - d)^2 + 4(h_i - h)^2}}{|d|}, \delta_e = \frac{\sqrt{\Delta d_e^2 + 4\Delta h_e^2}}{|d|} = \frac{\sqrt{(d_e - d)^2 + 4(h_e - h)^2}}{|d|}$$

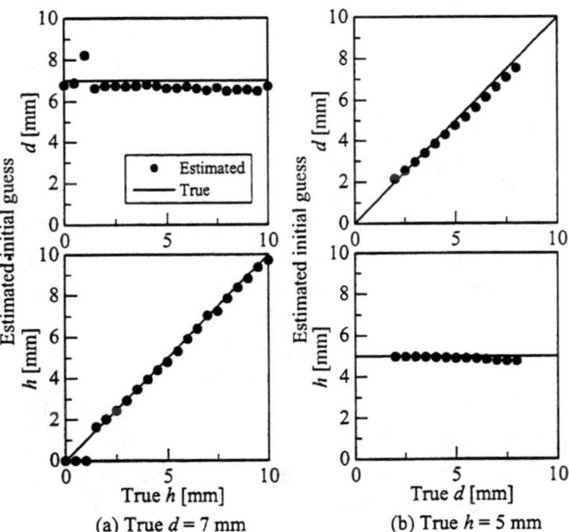


Figure 8. RESULT OF ROUGH ESTIMATION OF A FLAW

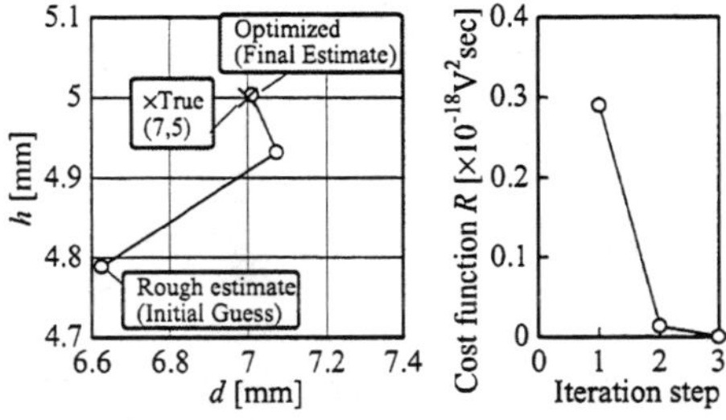


Figure 9. HISTORY OF OPTIMIZATION (IDENTIFICATION USING NUMERICAL RESULTS OF THE RECEIVER SIGNALS)

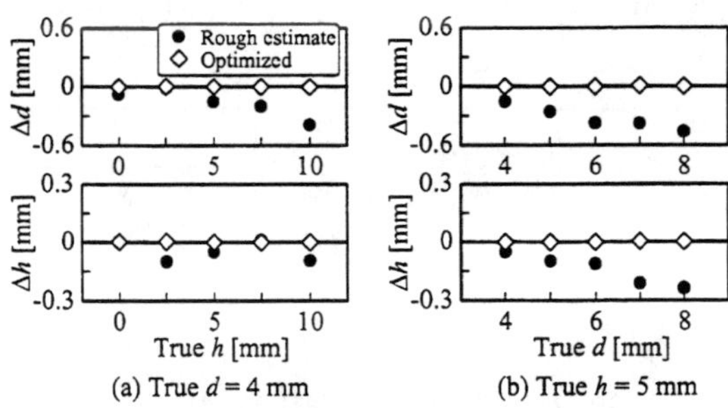


Figure 10. ERROR OF OPTIMIZED VALUES AND ROUGH ESTIMATE (IDENTIFICATION USING NUMERICAL RESULTS OF THE RECEIVER SIGNALS )

of their errors $\delta_i, \delta_e$. The result of the rough estimation of $h$ = 1mm, $d$ = 7mm is inaccurate; $\delta_i$ is greater than 30%. However, the identification using parameter optimization realizes very good result, since $\Delta h_e$, difference of the initial value of h from the true one, satisfy the condition mentioned above. Table 1 and Fig. 10 show that the results for various true values are accurate enough by using the identification method based on the inverse analysis.

**Identification Using Experimental Results of the Receiver Signals**

Table 2, Fig. 11 and Fig. 12 show the results of flaw identification using the receiver signals obtained by exper-

Table 2. MEASURED AND ESTIMATED VALUES (IDENTIFICATION USING EXPERIMENTAL RESULTS OF THE RECEIVER SIGNALS)

| Measured values | | Rough estimate | | Index of error | Optimized values | | Index of error |
|---|---|---|---|---|---|---|---|
| $h$ (mm) | $d$ (mm) | $h_i$ (mm) | $d_i$ (mm) | $\delta_i$ (%) | $h_e$ (mm) | $d_e$ (mm) | $\delta_e$ (%) |
| 0.004 6 | 4.030 0 | 0.000 0 | 4.065 9 | 0.920 2 | 0.000 9 | 4.070 7 | 1.026 5 |
| 2.359 5 | 4.074 6 | 2.429 4 | 4.177 9 | 4.265 8 | 2.521 0 | 4.189 5 | 8.414 8 |
| 4.909 6 | 4.011 2 | 4.935 8 | 3.763 6 | 6.309 7 | 5.001 9 | 4.010 9 | 4.604 3 |
| 7.358 1 | 4.070 2 | 7.433 8 | 3.861 0 | 6.345 3 | 7.393 1 | 4.267 2 | 5.136 2 |
| 9.928 2 | 4.016 8 | 9.765 6 | 3.909 6 | 8.524 6 | 9.898 3 | 4.216 6 | 5.192 7 |
| 4.999 4 | 5.095 7 | 4.851 4 | 4.978 6 | 6.247 7 | 5.024 7 | 5.184 3 | 2.002 4 |
| 4.874 0 | 6.009 2 | 4.750 3 | 5.752 6 | 5.932 8 | 4.875 7 | 6.264 3 | 4.245 4 |
| 4.855 6 | 7.023 0 | 4.600 9 | 6.667 8 | 8.842 2 | 4.794 2 | 7.238 3 | 3.529 7 |
| 4.866 8 | 8.126 1 | 4.737 6 | 7.696 5 | 6.169 2 | 4.949 4 | 8.231 3 | 2.409 7 |

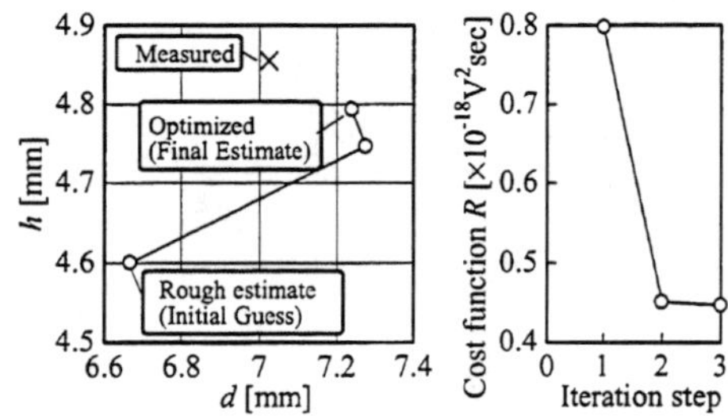


Figure 11. HISTORY OF OPTIMIZATION (IDENTIFICATION USING EXPERIMENTAL RESULTS OF THE RECEIVER SIGNALS)

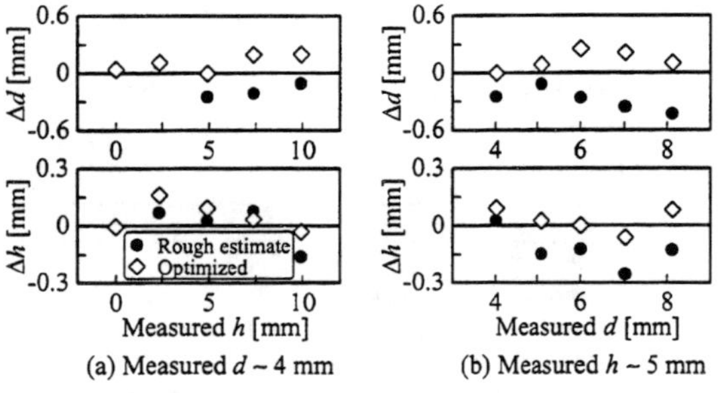


Figure 12. ERROR OF OPTIMIZED VALUES AND ROUGH ESTIMATE (IDENTIFICATION USING EXPERIMENTAL RESULTS OF THE RECEIVER SIGNALS )

iments. Measured parameters of the cavity were used as the standards for evaluating errors. Accuracy of the identification using optimization tends to be higher, though not strikingly, than that of the rough estimation. Thus, the inverse approach in this research can be effective also for the signal obtained from experiment.

## CONCLUSION

This research presented a method of quantitatively evaluating the parameters of a cavity modeling a flaw inside a specimen through inverse analysis of the receiver signals of an EMAT. Experimental results of the receiver signals agreed well with numerical ones, which verified the mathematical model of the inspection process. The flaw identification was formulated as a problem of parameter optimization. The condition that the initial guess must fulfill was discussed. A method of obtaining these initial values from the height and the time period of peaks in the receiver signals was also described. By applying the above method, flaw parameters were identified from the receiver signals obtained from numerical simulations and experiments, which verified the method of flaw identification presented here.

## REFERENCES


Hirose, S., 1993, "Inverse Scattering for Flaw Type Classification," Inverse Problems in Engineering Mechanics, Springer-Verlag, Berlin, pp.359-366.

Nishimura, N., 1994, "Numerical Solutions of Various Crack Determination Problems with BIEM," Lecture Notes in Num. Appl. Anal., Modelling, Computation and Analysis in Fracture Mechanics, **13**, pp.201-215.

Ludwig, R., You, Z., Palanisamy, R., 1993, "Numerical Simulation of Electromagnetic Acoustic Transducer-Receiver System for NDT Applications," IEEE Transaction on Magnetics, **29**-3, pp.2081-2089.

Sugiura, T., Maruyama, S., 2000, "Numerical Analysis of Ultrasonic Detection by an EMAT," Review of Progress in QNDE-2000, **20**, to be published.

Press, W. H., Teukolsky, S. A., Vetterling, W. T., Flannery, B. P., 1992, *Numerical Recipes in FORTRAN, 2nd ed.*, Cambrige University Press, pp.387-417.

Proceedings of
2001 ASME International Mechanical Engineering Congress and Exposition
November 11-16, 2001, New York, NY
NDE-Vol. 21

IMECE2001/NDE-25803

# ELASTIC WAVE PROPAGATION IN CIRCUMFERENTIAL DIRECTION IN ANISOTROPIC PIPES

S. Towfighi, T. Kundu* and M. Ehsani

Department of Civil Engineering and Engineering Mechanics
University of Arizona, Tucson, AZ 85721

*ASME Fellow, Member # 670711

## ABSTRACT

Ultrasonic nondestructive inspection of large diameter pipes is important for health monitoring of ailing infrastructure. Longitudinal stress-corrosion cracks are detected more efficiently by inducing circumferential waves; hence, the study of elastic wave propagation in the circumferential direction of a pipe is essential. The current state of knowledge lacks a complete solution of this problem. Only when the pipe material is isotropic a solution of the wave propagation problem in the circumferential direction exists. Ultrasonic inspections of reinforced concrete pipes and pipes retrofitted by fiber composites necessitate the development of a new theoretical solution for elastic wave propagation in anisotropic pipes in the circumferential direction. Mathematical modeling of the problem to obtain dispersion curves for anisotropic materials leads to coupled differential equations. Unlike isotropic materials for which the Stokes-Helmholtz decomposition technique simplifies the problem, in anisotropic case no such general decomposition technique works. These coupled differential equations are solved in this paper. Dispersion curves for anisotropic pipes of different curvatures have been computed and presented. Some numerical results computed by the new technique have been compared with those available in the literature.

## INTRODUCTION

Mathematical modeling of wave propagation in the axial direction of a cylinder has been studied extensively. However, for wave propagation in the circumferential direction, which is essential for Nondestructive Testing (NDT) of large diameter pipes, literature shows fewer investigations. Viktorov's work [1] establishes the fundamental mathematical modeling of the problem for isotropic material properties. He has introduced the angular wave number concept and has derived, decomposed and solved the governing differential equations. He has considered only one curved surface; in other words, he has found the solution for convex and concave cylindrical surfaces. In order to obtain the results for pipes Qu et al. [2] have added the boundary conditions for the second surface and solved the problem of guided wave propagation in isotropic curved plates.

Unlike isotropic materials for which the Stokes-Helmholtz decomposition technique simplifies the problem, for anisotropic case no such general decomposition technique works. The differential equations remain coupled and require a more general solution technique.

The new technique, presented in this paper, solves coupled set of differential equations without attempting to decopule the equations. Hence it removes the obstacle arising from not being able to decouple the equations. Consequently it provides a systematic and unifying solution method, which is capable of solving a set of coupled differential equations, and can be utilized to solve a variety of wave propagation problems.

## FUNDAMENTAL EQUATIONS

Wave propagation in circumferential direction in pipes with isotropic material properties is usually modeled as a plane strain problem; i.e. the displacement component along the longitudinal axis of the pipe is set equal to zero. For a few other types of anisotropy this situation remains valid. However, for general anisotropy the longitudinal component of displacement must be considered in the mathematical modeling. The symmetry of both geometry and material properties is required for plane strain idealization. In absence of such symmetry a three-dimensional mathematical modeling is necessary.

In cylindrical coordinates, strain components in terms of displacements can be written as:

$$e_{rr} = \frac{\partial u_r(r,\theta,t)}{\partial r} \tag{1}$$

$$e_{\theta\theta} = \frac{\partial u_\theta(r,\theta,t)}{r\partial\theta} + \frac{1}{r}u_r(r,\theta,t)$$

$$e_{zz} = \frac{\partial u_z(r,\theta,t)}{\partial z}$$

$$e_{r\theta} = \frac{1}{2}\left(\frac{\partial u_r(r,\theta,t)}{r\partial\theta} + \frac{\partial u_\theta(r,\theta,t)}{\partial r} - \frac{u_\theta(r,\theta,t)}{r}\right)$$

$$e_{\theta z} = \frac{1}{2}\left(\frac{\partial u_z(r,\theta,t)}{r\partial\theta} + \frac{\partial u_\theta(r,\theta,t)}{\partial z}\right)$$

$$e_{rz} = \frac{1}{2}\left(\frac{\partial u_r(r,\theta,t)}{\partial z} + \frac{\partial u_z(r,\theta,t)}{\partial r}\right)$$

The stress and displacement components are shown in Fig. 1.

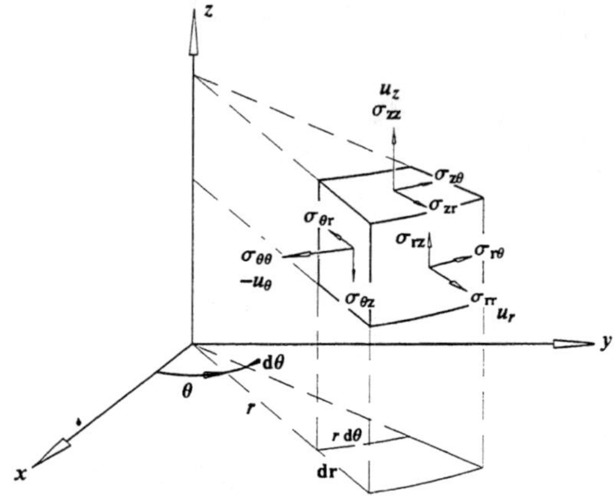


*Fig. 1 – Stress and Displacement components in cylindrical coordinate system*

And constitutive matrix for general anisotropy contains 21 independent elastic constants:

(2)

$$\begin{pmatrix} \sigma_{\theta,\theta} \\ \sigma_{z,z} \\ \sigma_{r,r} \\ \sigma_{\theta,z} \\ \sigma_{r,\theta} \\ \sigma_{r,z} \end{pmatrix} = \begin{pmatrix} C_{1,1} & C_{1,2} & C_{1,3} & C_{1,4} & C_{1,5} & C_{1,6} \\ C_{1,2} & C_{2,2} & C_{2,3} & C_{2,4} & C_{2,5} & C_{2,6} \\ C_{1,3} & C_{2,3} & C_{3,3} & C_{3,4} & C_{3,5} & C_{3,6} \\ C_{1,4} & C_{2,4} & C_{3,4} & C_{4,4} & C_{4,5} & C_{4,6} \\ C_{1,5} & C_{2,5} & C_{3,5} & C_{4,5} & C_{5,5} & C_{5,6} \\ C_{1,6} & C_{2,6} & C_{3,6} & C_{4,6} & C_{5,6} & C_{6,6} \end{pmatrix} \cdot \begin{pmatrix} e_{\theta,\theta} \\ e_{z,z} \\ e_{r,r} \\ 2\,e_{\theta,z} \\ 2\,e_{r,\theta} \\ 2\,e_{r,z} \end{pmatrix}$$

Equations of motion for three components of displacements in cylindrical coordinates are as follows:

(3)

$$\frac{\partial \sigma_{rr}}{\partial r} + \frac{\partial \sigma_{rz}}{\partial z} + \frac{\partial \sigma_{r\theta}}{r\partial \theta} + \frac{\sigma_{rr} - \sigma_{\theta\theta}}{r} - \rho \frac{\partial^2 u_r(r, \theta, t)}{\partial t^2} = 0$$

$$\frac{\partial \sigma_{r\theta}}{\partial r} + \frac{\partial \sigma_{\theta z}}{\partial z} + \frac{\partial \sigma_{\theta\theta}}{r\partial \theta} + \frac{2\,\sigma_{r\theta}}{r} - \rho \frac{\partial^2 u_\theta(r, \theta, t)}{\partial t^2} = 0$$

$$\frac{\partial \sigma_{rz}}{\partial r} + \frac{\partial \sigma_{zz}}{\partial z} + \frac{\partial \sigma_{\theta z}}{r\partial \theta} + \frac{\sigma_{rz}}{r} - \rho \frac{\partial^2 u_z(r, \theta, t)}{\partial t^2} = 0$$

Stress components in the above equations can be substituted in terms of displacement components. Since displacement components are functions of wave forms, time dependency of waves must be established.

## WAVE FORM

In cylindrical geometry the generation of surface waves in the circumferential direction with a plane wave front requires the circumferential wave speed to be a function of the radial distance. Viktorov [1] has introduced this concept and called it the angular wave number. Similar formulation has been adapted here:

(4)

$$u_r(r, \theta, t) = U_r(r)\, e^{i\,(p\theta - \omega t)}$$

$$u_\theta(r, \theta, t) = U_t(r)\, e^{i\,(p\theta - \omega t)}$$

$$u_z(r, \theta, t) = U_z(r)\, e^{i\,(p\theta - \omega t)}$$

where $U_r(r)$ $U_t(r)$ and $U_z(r)$ represent the amplitude of vibration in the radial, tangential, and axial directions, respectively. "i" is the imaginary number $\sqrt{-1}$ and p is the angular wave number equal to the ratio of angular velocity to phase velocity times the radius. It should be noted here that the phase velocity is not a constant and changes with radius. Hence, if "c" is assumed to be the phase velocity at the outer surface with radius b; for other points having a radius r the phase velocity would be $cr/b$ .

## GOVERNING DIFFERENTIAL EQUATIONS

Subsequent substitution of equations (4), (1) and (2) into equations (3) yields the following governing differential equations:

(5)

$$\begin{aligned} &-2\,C_{5,5}\,U_r(r)\,p^2 - 2\,C_{1,5}\,U_t(r)\,p^2 - 2\,C_{4,5}\,U_z(r)\,p^2 - 2\,i\,C_{1,1}\,U_t(r)\,p - \\ &2\,i\,C_{5,5}\,U_t(r)\,p - 2\,i\,C_{1,4}\,U_z(r)\,p + 4\,i\,r\,C_{3,5}\,U_r'(r)\,p + \\ &2\,i\,r\,C_{1,3}\,U_t'(r)\,p + 2\,i\,r\,C_{5,5}\,U_t'(r)\,p + 2\,i\,r\,C_{3,4}\,U_z'(r)\,p + \\ &2\,i\,r\,C_{5,6}\,U_z'(r)\,p + 2\,r^2\,\rho\,\omega^2\,U_r(r) - 2\,C_{1,1}\,U_r(r) + 2\,C_{1,5}\,U_t(r) + \\ &2\,r\,C_{3,3}\,U_r'(r) - 2\,r\,C_{1,5}\,U_t'(r) - 2\,r\,C_{1,6}\,U_z'(r) + 2\,r\,C_{3,6}\,U_z'(r) + \\ &2\,r^2\,C_{3,3}\,U_r''(r) + 2\,r^2\,C_{3,5}\,U_t''(r) + 2\,r^2\,C_{3,6}\,U_z''(r) = 0 \end{aligned}$$

$$\begin{aligned} &-2\,C_{1,5}\,U_r(r)\,p^2 - 2\,C_{1,1}\,U_t(r)\,p^2 - 2\,C_{1,4}\,U_z(r)\,p^2 + 2\,i\,C_{1,1}\,U_r(r)\,p + \\ &2\,i\,C_{5,5}\,U_r(r)\,p + 2\,i\,C_{4,5}\,U_z(r)\,p + 2\,i\,r\,C_{1,3}\,U_r'(r)\,p + 2\,i\,r\,C_{5,5}\,U_r'(r)\,p + \\ &4\,i\,r\,C_{1,5}\,U_t'(r)\,p + 2\,i\,r\,C_{1,6}\,U_z'(r)\,p + 2\,i\,r\,C_{4,5}\,U_z'(r)\,p + 2\,C_{1,5}\,U_r(r) + \\ &2\,r^2\,\rho\,\omega^2\,U_t(r) - 2\,C_{5,5}\,U_t(r) + 2\,r\,C_{1,5}\,U_r'(r) + 4\,r\,C_{3,5}\,U_r'(r) + \\ &2\,r\,C_{5,5}\,U_t'(r) + 4\,r\,C_{5,6}\,U_z'(r) + 2\,r^2\,C_{3,5}\,U_r''(r) + 2\,r^2\,C_{5,5}\,U_t''(r) + \\ &2\,r^2\,C_{5,6}\,U_z''(r) = 0 \end{aligned}$$

$$\begin{aligned} &-2\,C_{4,5}\,U_r(r)\,p^2 - 2\,C_{1,4}\,U_t(r)\,p^2 - 2\,C_{4,4}\,U_z(r)\,p^2 + 2\,i\,C_{1,4}\,U_r(r)\,p - \\ &2\,i\,C_{4,5}\,U_t(r)\,p + 2\,i\,r\,C_{3,4}\,U_r'(r)\,p + 2\,i\,r\,C_{5,6}\,U_r'(r)\,p + \\ &2\,i\,r\,C_{1,6}\,U_t'(r)\,p + 2\,i\,r\,C_{4,5}\,U_t'(r)\,p + 4\,i\,r\,C_{4,6}\,U_z'(r)\,p + \\ &2\,r^2\,\rho\,\omega^2\,U_z(r) + 2\,r\,C_{1,6}\,U_r'(r) + 2\,r\,C_{3,6}\,U_r'(r) + 2\,r\,C_{6,6}\,U_z'(r) + \\ &2\,r^2\,C_{3,6}\,U_r''(r) + 2\,r^2\,C_{5,6}\,U_t''(r) + 2\,r^2\,C_{6,6}\,U_z''(r) = 0 \end{aligned}$$

## BOUNDARY CONDITIONS

In order to obtain the dispersion curves, the traction free boundary conditions (zero stress values on the inner and outer surfaces of the pipe) must be satisfied.
Hence, at r = a and r = b:

(6)

$$\begin{aligned} &C_{1,3}\,U_r(r) + i\,p\,C_{3,5}\,U_r(r) + i\,p\,C_{1,3}\,U_t(r) - C_{3,5}\,U_t(r) + i\,p\,C_{3,4}\,U_z(r) + \\ &r\,C_{3,3}\,U_r'(r) + r\,C_{3,5}\,U_t'(r) + r\,C_{3,6}\,U_z'(r) = 0 \end{aligned}$$

$$\begin{aligned} &C_{1,5}\,U_r(r) + i\,p\,C_{5,5}\,U_r(r) + i\,p\,C_{1,5}\,U_t(r) - C_{5,5}\,U_t(r) + i\,p\,C_{4,5}\,U_z(r) + \\ &r\,C_{3,5}\,U_r'(r) + r\,C_{5,5}\,U_t'(r) + r\,C_{5,6}\,U_z'(r) = 0 \end{aligned}$$

$$C_{1,6}\,U_r(r) + i\,p\,C_{5,6}\,U_r(r) + i\,p\,C_{1,6}\,U_t(r) - C_{5,6}\,U_t(r) +$$
$$i\,p\,C_{4,6}\,U_z(r) + r\,C_{3,6}\,U_r'(r) + r\,C_{5,6}\,U_t'(r) + r\,C_{6,6}\,U_z'(r) = 0$$

## SOLUTION

It can be seen that all differential equations are functions of three displacement components $U_r(r)$, $U_t(r)$, $U_z(r)$ and their derivatives. It should be also noted that $U_r(r)$, $U_t(r)$ and $U_z(r)$ are functions of the radius only and they appear in all equations. Therefore, there are three coupled differential equations and six boundary conditions that must be satisfied simultaneously.

To solve the equations, the unknown functions are expanded in Fourier Series (FS). Substitution of FS Expansions into the differential equations provides three algebraic equations that must be satisfied for the entire problem domain. To satisfy the equations for a given number of FS terms weighted residuals integration with a linear weight function has been utilized:

$$R = \int_a^b w\, f(r,\, x_i)\, dr = 0 \tag{7}$$

The radius corresponding to the peak value of the linear weight function can take any value between the inner and the outer radius, each resulting one independent equation. Hence from every differential equation any number of equations can be obtained.

On the other hand, it is known that the general solution is a linear combination of all solution functions that can be obtained. Therefore, the general solution should contain combinatorial parameters. The number of combinatorial parameters is the same as the number of individual solutions. These combinatorial parameters are necessary to satisfy the boundary conditions. Satisfaction of six boundary conditions requires six parameters and six equations. Therefore the necessary and sufficient number of combinatorial parameters is six and it indicates the existence of six independent solutions.

Substitution of solution functions into the differential equations leads to three equations, each containing all of the FS parameters. In other words, all FS parameters for the three amplitude functions appear in every equation. Because of this coupling, the values of parameters obtained for FS expansion of $U_r(r)$, $U_t(r)$ and $U_z(r)$ are not independent and a solution must yield all parameters as one set of results. Since the equations are linear and the results must be combined using combinatorial parameters only their relative values must be found. Therefore one of the FS parameters can be assumed equal to one. Then the relative values for other FS parameters can be calculated in terms of this unit value. Each set of the parameter values defines a set of dependent shapes for the above amplitude functions; these are called basic shapes. Since the number of equations must be equal to the number of unknowns a specific number of weight functions are required.

The FS expansion for $U_r(r)$ can be written as:

$$U_r(r) = x_0 + \sum_{n=1}^{m} \left( \cos\!\left(\frac{n\pi r}{L}\right) x_n + \sin\!\left(\frac{n\pi r}{L}\right) y_n \right) \tag{8}$$

which contains $2m+1$ parameters or coefficients, $x_n$ and $y_n$.

With two other expressions for $U_t(r)$ and $U_z(r)$ the number of unknowns increases to $6m+3$. Performing weighted residuals method, a set of linear equations results:

$$\begin{pmatrix} a_{1,1}x_1 & a_{1,2}x_2 & \cdots & a_{1,s}x_s & a_{1,s+1}x_{s+1} & \cdot & a_{1,s+6}x_{s+6} \\ a_{2,1}x_1 & a_{2,2}x_2 & \cdots & a_{2,s}x_s & a_{2,s+1}x_{s+1} & \cdot & a_{2,s+6}x_{s+6} \\ \cdot & \cdot & \cdots & \cdot & \cdot & \cdot & \cdot \\ \cdot & \cdot & \cdots & \cdot & \cdot & \cdot & \cdot \\ a_{s,1}x_1 & a_{s,2}x_2 & \cdots & a_{s,s}x_s & a_{s,s+1}x_{s+1} & \cdot & a_{s,s+6}x_{s+6} \end{pmatrix} = \begin{pmatrix} 0 \\ 0 \\ \cdot \\ \cdot \\ 0 \end{pmatrix} \tag{9}$$

where $x_{s+1}, x_{s+2}, \ldots, x_{s+6}$ represent the last sine and cosine terms of FS expansions. Assigning six independent unit vectors to the last six parameters as shown in Equation (10),

$$\begin{pmatrix} x^1_{s+1} & x^2_{s+1} & x^3_{s+1} & x^4_{s+1} & x^5_{s+1} & x^6_{s+1} \\ x^1_{s+2} & x^2_{s+2} & x^3_{s+2} & x^4_{s+2} & x^5_{s+2} & x^6_{s+2} \\ x^1_{s+3} & x^2_{s+3} & x^3_{s+3} & x^4_{s+3} & x^5_{s+3} & x^6_{s+3} \\ x^1_{s+4} & x^2_{s+4} & x^3_{s+4} & x^4_{s+4} & x^5_{s+4} & x^6_{s+4} \\ x^1_{s+5} & x^2_{s+5} & x^3_{s+5} & x^4_{s+5} & x^5_{s+5} & x^6_{s+5} \\ x^1_{s+6} & x^2_{s+6} & x^3_{s+6} & x^4_{s+6} & x^5_{s+6} & x^6_{s+6} \end{pmatrix} = \begin{pmatrix} 1 & 0 & 0 & 0 & 0 & 0 \\ 0 & 1 & 0 & 0 & 0 & 0 \\ 0 & 0 & 1 & 0 & 0 & 0 \\ 0 & 0 & 0 & 1 & 0 & 0 \\ 0 & 0 & 0 & 0 & 1 & 0 \\ 0 & 0 & 0 & 0 & 0 & 1 \end{pmatrix} \tag{10}$$

yields six independent solutions. Therefore the number of equations has to be $s = 6m - 3$. Consequently, the general solution can be obtained as a linear combination of the above solutions.

$$A_1 \begin{pmatrix} x^1_1 \\ x^1_2 \\ \cdot \\ \cdot \\ \cdot \\ x^1_s \end{pmatrix} + A_2 \begin{pmatrix} x^2_1 \\ x^2_2 \\ \cdot \\ \cdot \\ \cdot \\ x^2_s \end{pmatrix} + A_3 \begin{pmatrix} x^3_1 \\ x^3_2 \\ \cdot \\ \cdot \\ \cdot \\ x^3_s \end{pmatrix} + A_4 \begin{pmatrix} x^4_1 \\ x^4_2 \\ \cdot \\ \cdot \\ \cdot \\ x^4_s \end{pmatrix} + A_5 \begin{pmatrix} x^5_1 \\ x^5_2 \\ \cdot \\ \cdot \\ \cdot \\ x^5_s \end{pmatrix} + A_6 \begin{pmatrix} x^6_1 \\ x^6_2 \\ \cdot \\ \cdot \\ \cdot \\ x^6_s \end{pmatrix} \tag{11}$$

Superscript for FS parameters shows the solution set number.

Substitution of the obtained FS parameters into stress components on the inner and outer surfaces of the pipe leads to an eigenvalue problem. The determinant of the coefficients of $A_i$ for any point located on the dispersion curves should be zero.

## NUMERICAL RESULTS

Based on the proposed mathematical modeling a Mathematica program has been developed. To ensure the validity of the modeling and the computer program, its results are compared with the available dispersion curves for anisotropic flat plates by using small ratios of thickness to radius, when pipe geometry approaches flat plate geometry. Additionally, the results are compared with the published results for isotropic pipes [Qu et al. (1996)]. Since the exact input values have not been reported by Qu et al. (1996), the comparison is done only qualitatively. The dispersion curves are also given for anisotropic pipes.

**A) Comparison with available data for isotropic flat plate**

Dispersion curves for a flat plate are given in Mal and Singh [3], see Fig. 2. Curves for the same plate thickness and material properties but having a radius of curvature of 1 m are generated by the proposed method and shown in Fig. 3.

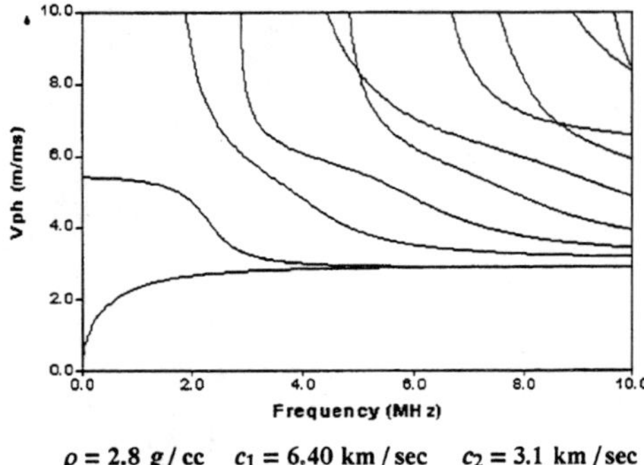


$\rho = 2.8\ g/cc \quad c_1 = 6.40\ \mathrm{km/sec} \quad c_2 = 3.1\ \mathrm{km/sec}$

*Fig. 2 – Dispersion curves for isotropic flat plate [Mal and Singh (1991)]. Plate thickness = 1 mm.*

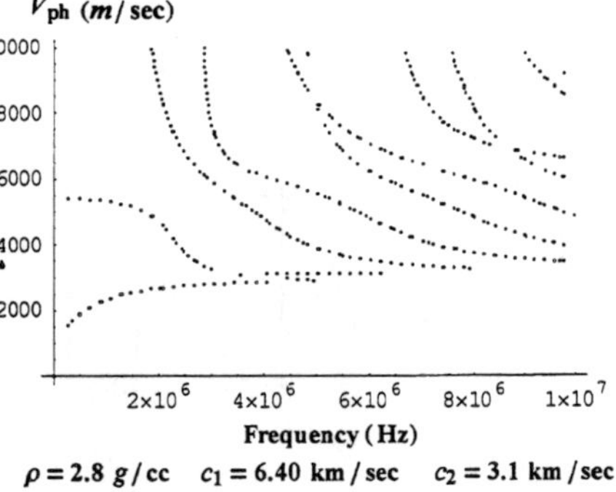


$\rho = 2.8\ g/cc \quad c_1 = 6.40\ \mathrm{km/sec} \quad c_2 = 3.1\ \mathrm{km/sec}$

*Fig. 3 – Dispersion curves generated by the proposed method. Plate thickness = 1 mm. Pipe outside radius = 1.0 m*

A comparison of Figures 2 and 3 shows a very good match between the two when only 20 terms are used in the FS Expansions.

**B) Comparison with available data for anisotropic flat plate**

For anisotropic flat plates, dispersion curves are given in Rose [4]. For the unidirectional composite plate or pipe with a zero degree angle between the wave propagation direction and the fiber direction as shown in Fig. 4, the material and the geometric symmetry conditions are maintained; hence, the plain strain formulation remains valid. Consequently the constitutive matrix reduces to the following form,

$$\begin{pmatrix} \sigma_{\theta\theta} \\ \sigma_{zz} \\ \sigma_{rr} \\ \sigma_{r\theta} \end{pmatrix} = \begin{pmatrix} 128.2 & 6.9 & 6.9 & 0 \\ 6.9 & 14.95 & 7.33 & 0 \\ 6.9 & 7.33 & 14.95 & 0 \\ 0 & 0 & 0 & 6.73 \end{pmatrix} \begin{pmatrix} e_{\theta\theta} \\ 0 \\ e_{rr} \\ 2e_{r\theta} \end{pmatrix} \qquad (12)$$

Stiffness values are given in GPa.

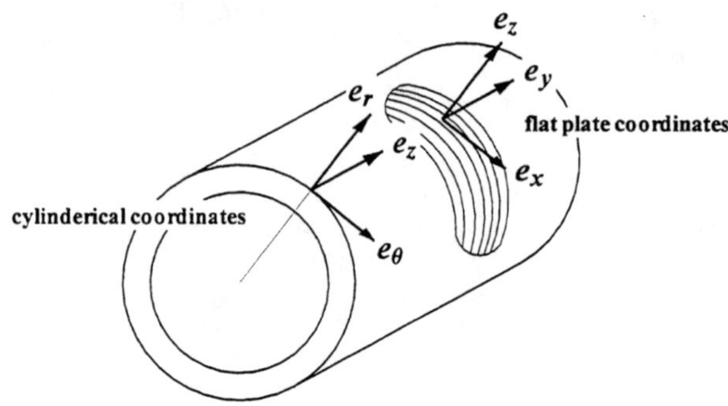


*Fig. 4 – Tangential direction of the fibers maintains the symmetry. Coordinate systems for flat plate and pipe analyses are also shown.*

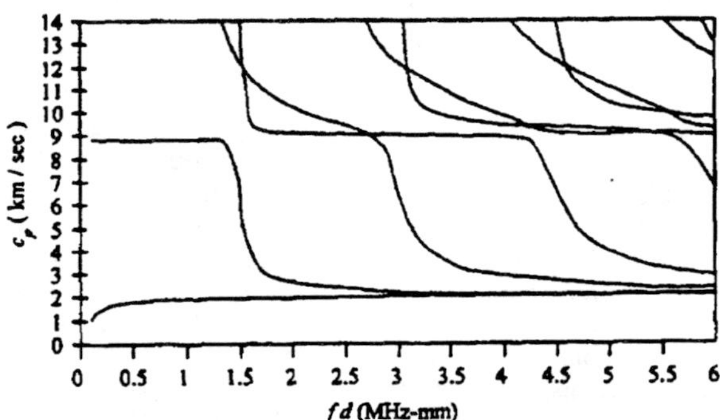


*Fig. 5 – Dispersion curves of a unidirectional composite plate for waves propagating in fiber direction (x-axis direction, 0°). Material properties are given in Eq.(12), $\rho = 1580\ \mathrm{kg/m^3}$ [after Rose(1999)].*

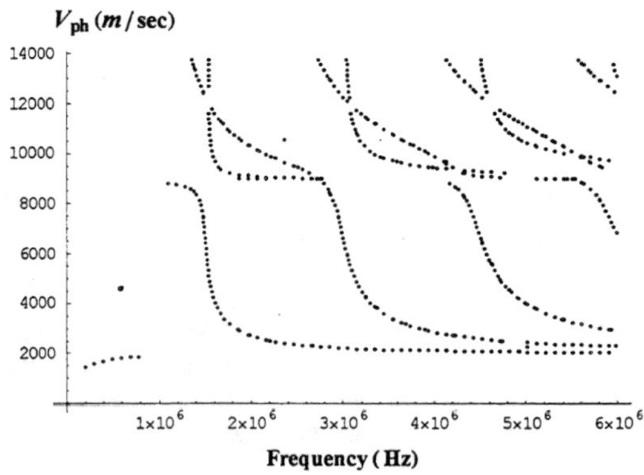


*Fig. 6 – Dispersion curves for a large diameter pipe made of an anisotropic material. Material properties are given in Eq.(12). Pipe wall thickness = 1mm. Pipe radius = 1000 mm, m = 30.*

The result of Fig. 6 is obtained using 30 terms (m=30) in the Fourier Series Expansion. To show the effect of the number of terms (m) on the computed results the same dispersion curves are computed for m = 20 and shown in Fig. 7.

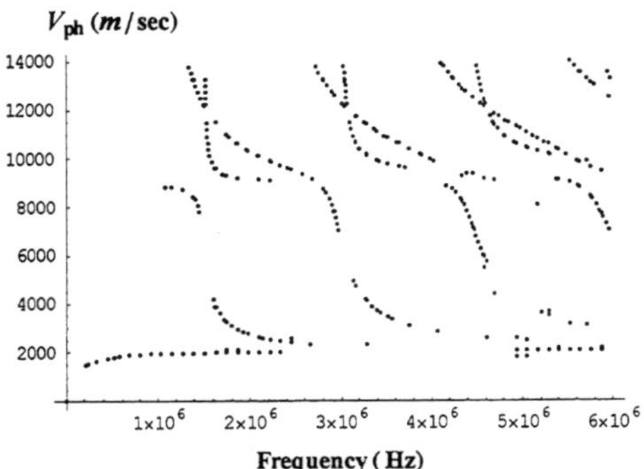


*Fig. 7 – Dispersion curves for the anisotropic pipe with m=20. Pipe dimensions and material properties are same as in Fig.6, only m is different.*

It is interesting to note that smaller value of m gives broken lines but not erroneous results. Therefore the user can easily realize the need for a greater number of terms in the FS expansion when the lines in the dispersion curve plot are found broken. There are some missing parts of curves in Fig.6 that can be obtained by increasing m. However for m = 30 we get enough information for comparison with the results given by Rose (1999).

For the same material with fibers going in the longitudinal direction of the pipe, the constitutive matrix changes to Eq.(13).

$$\begin{pmatrix} \sigma_{\theta\theta} \\ \sigma_{zz} \\ \sigma_{rr} \\ \sigma_{r\theta} \end{pmatrix} = \begin{pmatrix} 14.95 & 6.9 & 7.33 & 0 \\ 6.9 & 128.2 & 6.9 & 0 \\ 7.33 & 6.9 & 14.95 & 0 \\ 0 & 0 & 0 & 3.81 \end{pmatrix} \begin{pmatrix} e_{\theta\theta} \\ 0 \\ e_{rr} \\ 2\,e_{r\theta} \end{pmatrix} \tag{13}$$

Obtained results for this case also match with the corresponding dispersion curves presented by Rose (1999), See figures 8 and 9.

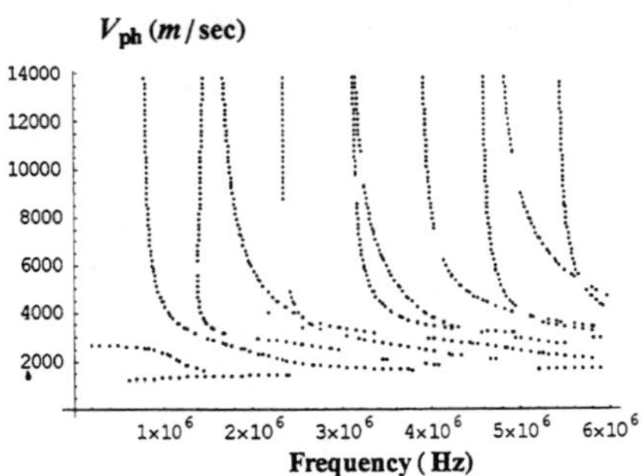


*Fig. 8 – Computed dispersion curves for an anisotropic large diameter pipe, when fiber and wave propagation directions are perpendicular to each other. Material properties are given in Eq.(13). Pipe wall thickness = 1 mm. Pipe radius = 1000 mm.*

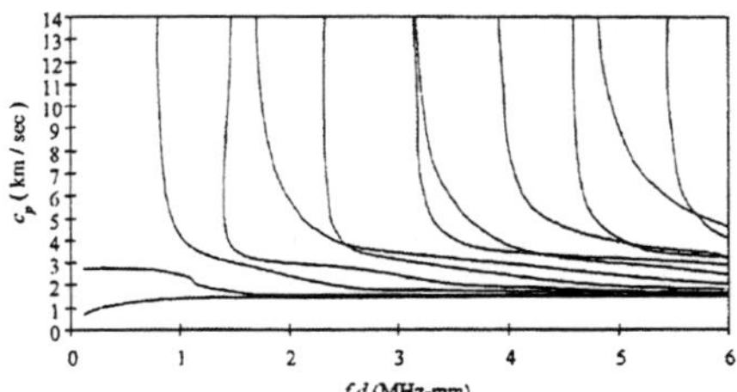


*Fig. 9 – Dispersion curves of unidirectional composite plate for waves propagating perpendicular to the fiber direction (x-axis direction, 90 °). Material properties are given in Eq.(13). Plate thickness = 1 mm, $\rho = 1580$ kg/m³ [Rose, 1999].*

## C) Comparison with available data for isotropic pipe

As mentioned earlier Qu et al. (1996) have derived dispersion curves for aluminum pipes but the material properties have not been reported in their work. Hence, the quantitative comparison was not possible. However, curves presented here, qualitatively look similar to those of Qu et al. (1996). Fig. 8 shows the obtained dispersion curves with non-dimensional $\bar{k}$ and $\bar{\omega}$ where $\bar{k} = k(b-a)$ and $\bar{\omega} = \omega(b-a)\sqrt{\dfrac{\rho}{\mu}}$.

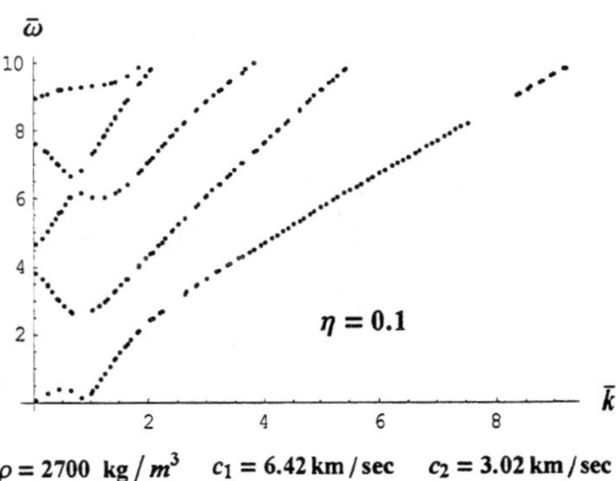


$\rho = 2700$ kg/m³ $\quad c_1 = 6.42$ km/sec $\quad c_2 = 3.02$ km/sec

*Fig. 10 – Dispersion curves for aluminum pipe obtained by the proposed method. $\eta$ = ratio of inner to outer radius = 0.1*

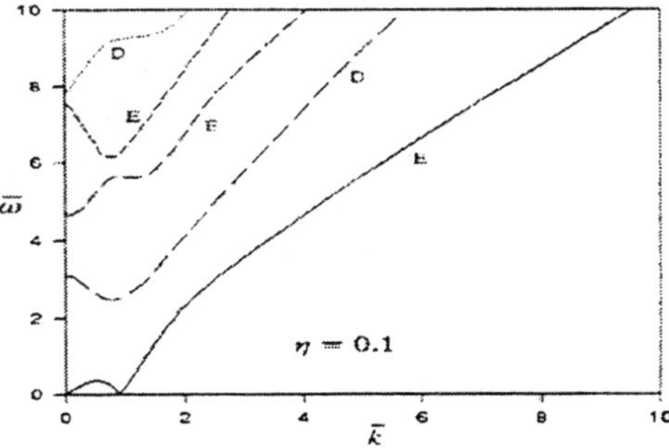


*Fig. 11 – Dispersion curves for aluminum pipe obtained by Qu et al.(1996). Material properties are not known.*

## D) Anisotropic Pipe of Small Radius of Curvature

To show the effect of the radius of curvature on the dispersion curves the pipe radius is varied from 1000 mm to 2.5 mm keeping the wall thickness and material properties same as those mentioned in the figure captions for Figures 6 and 8. Dispersion curves obtained by the 30 terms FS expansion for r = 1000, 10, 5 and 2.5 mm are shown in Figures 12 and 13. Figure 12 shows dispersion curves for fibers going in the circumferential direction and Figure 13 is for fibers going in the axial direction while the waves propagate in the circumferential directions in both cases.

From Figure 12 one can see that for fibers oriented in the circumferential direction the dispersion curves do not change significantly as the radius of curvature (r) is reduced from 1000 mm to 10 mm. However, as r is reduced further the deviation of the dispersion curves from the large radius case is no longer negligible. For fibers oriented in the axial direction (Figure 13) the dispersion curves remain almost unchanged for r = 1000 mm to 5 mm. However, as it is reduced further (r = 2.5 mm) significant changes occur. For r = 2.5 mm the dispersion curves show many broken lines. To get continuous lines more terms in the Fourier Series expansion should be considered.

In summary, a comparison between Figures 12 and 13 shows that the effect of curvature is stronger when the fibers are oriented along the circumferential direction. In other words, for fibers oriented in the axial direction the flat plate approximation can be extended to pipes of lower radius or higher curvature.

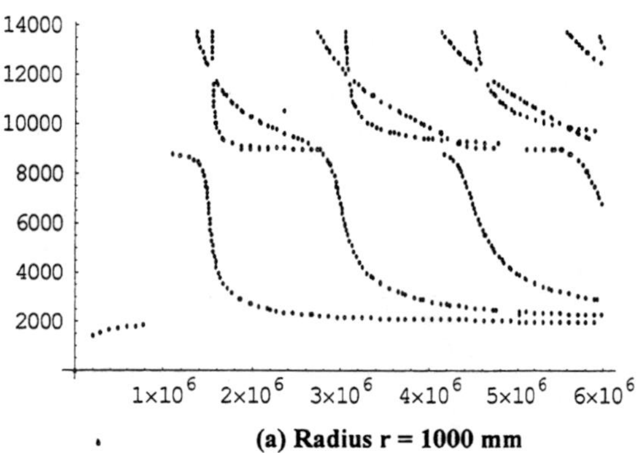


(a) Radius r = 1000 mm

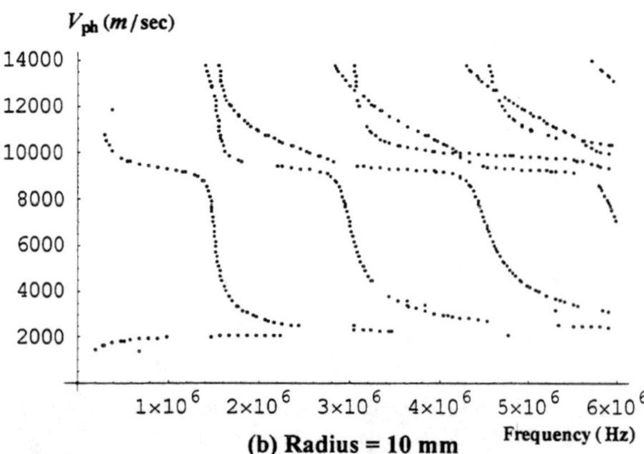


(b) Radius = 10 mm

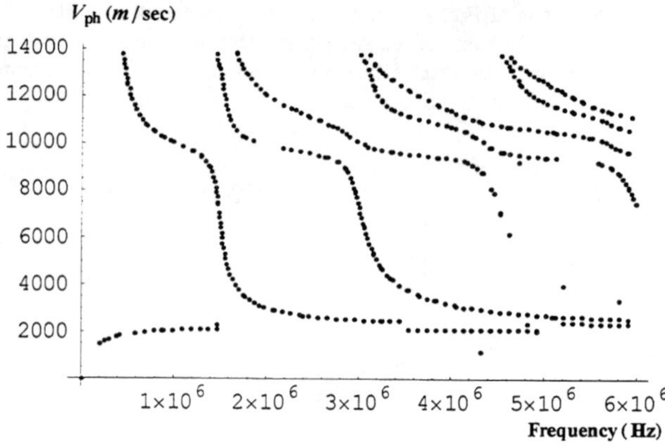


(c) Radius = 5 mm

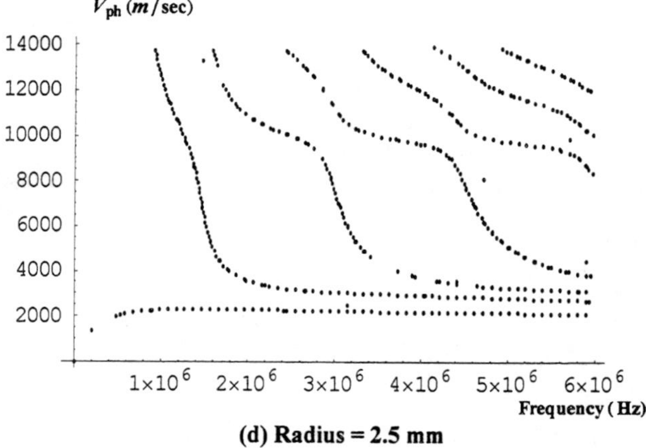


(d) Radius = 2.5 mm

*Figure 12 - Dispersion curves for circumferential direction wave propagation in fiber reinforced composite pipes when fibers are oriented in the circumferential direction, radius of curvature of the pipe is (a) 1000 mm, (b) 10 mm, (c) 5 mm (next page), and (d) 2.5 mm (next page). Pipe wall thickness and material properties are same as those in Figure 6.*

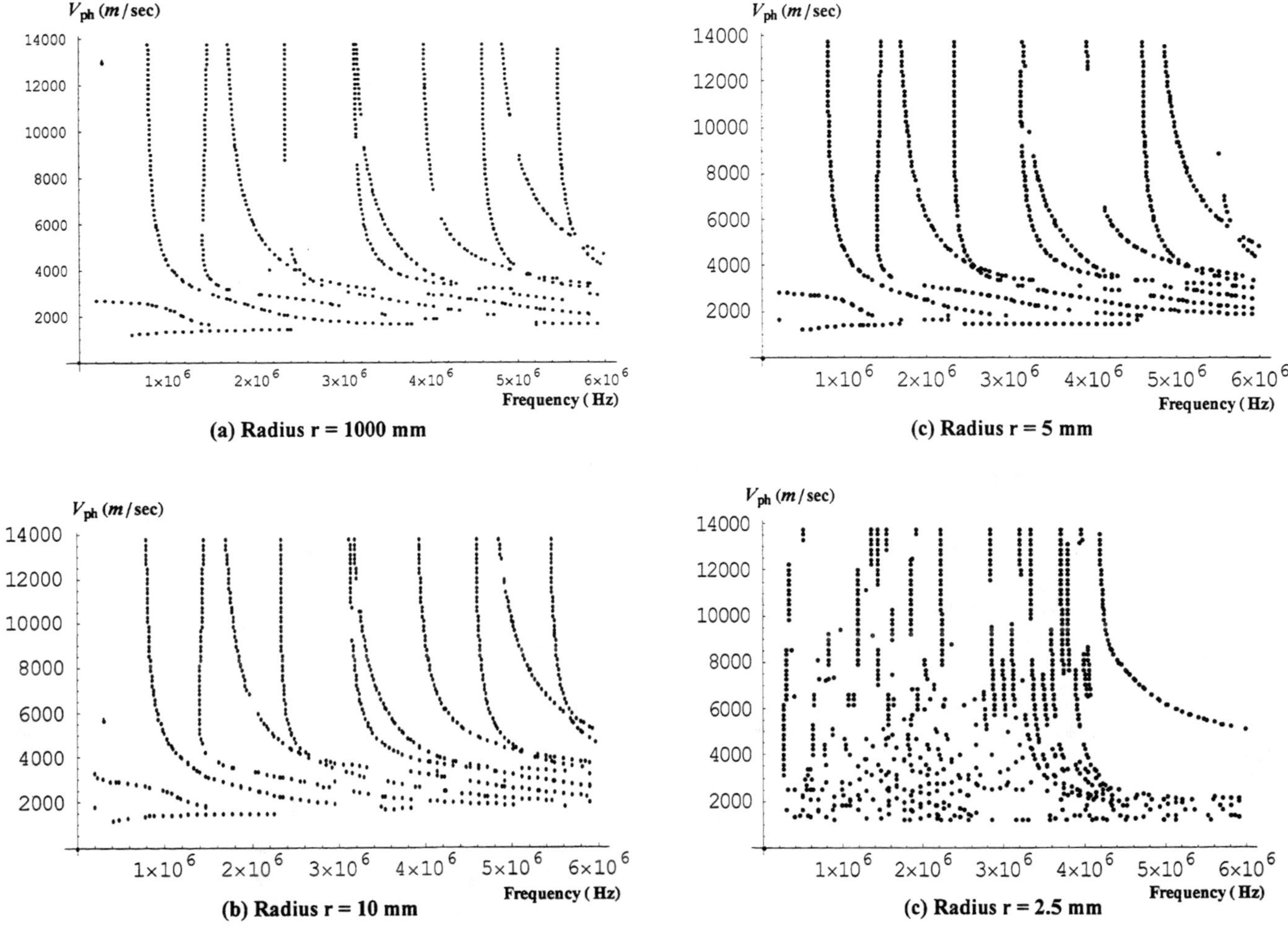


*Figure 13 - Dispersion curves for circumferential direction wave propagation in fiber reinforced composite pipes when fibers are oriented in the axial direction, radius of curvature of the pipe is (a) 1000 mm, (b) 10 mm, (c) 5 mm (next page), and (d) 2.5 mm (next page). Pipe wall thickness and material properties are same as those in Figure 8.*

## CONCLUSION

A solution technique based on the Fourier series expansion of the unknown quantities has been introduced to solve the elastic wave propagation problem in anisotropic cylinders in the circumferential direction. Accuracy of the technique has been verified by comparing the computed results for isotropic pipes with the published results. Since no published results are available for wave propagation in the circumferential direction in anistropic pipes, the computed dispersion curves for anisotropic pipes could not be compared with any results available in the literature. However, the Lamb wave dispersion curves for flat plates can be computed and those values are used to check the accuracy of the proposed technique. With the new technique dispersion curves for pipes with large diameter (diameter to thickness ratio is 2000) have been computed and compared with the flat plate results for both isotropic and anisotropic materials.

Computed results for such low curvature pipes matched very well with the flat plate results. The effect on the dispersion curves as the curvature of the anisotropic plate increases has been also studied here.

The solution technique used to solve this specific wave propagation problem is a newly developed general solution technique for solving a coupled partial differential equation set [Towfighi, (2001)]. Applicability of this technique to other wave propagation problems is currently under investigation.

## ACKNOWLEDGEMENT

This research was partially supported from an NSF grant CMS-9901221.

## REFERENCES

1. Viktorov, I. A., 1958, "Rayleigh-type Waves on a Cylindrical Surface," Soviet Physics – Acoustics, vol. 4, pp. 131-136, 1958

2. Qu, J., Berthelot, Y., and Li, Z. 1996, "Dispersion of guided circumferential waves in a circular annulus," In D. O. Thompson and D. E. Chimenti (Eds.). Review of Progress in Quantitative Nondestructive Evaluation, vol. 15, pp. 169-76. New York: Plenum

3. Mal, A. K., and Singh., S. J. 1991, *Deformation of Elastic Solids*, Prentice-Hall Inc., p. 313.

4. Rose, J. L., 1999, "Ultrasonic Waves in Solid Media," Cambridge University Press, pp. 264-271.

5. Towfighi, S., 2001, "Elastic Wave Propagation in Circumferential Direction in Anisotropic Pipes," Ph.D. Dissertation, 2001, The University of Arizona, Tucson, AZ.

Proceedings of
2001 ASME International Mechanical Engineering Congress and Exposition
November 11–16, 2001, New York, NY
NDE-Vol. 21

IMECE2001/NDE-25804

# UNDERWATER PIPELINE INSPECTION USING GUIDED WAVES

Won-Bae Na and Tribikram Kundu

Department of Civil Engineering and Engineering Mechanics

The University of Arizona, Tucson, AZ 85721, USA

## ABSTRACT

Underwater pipeline inspections are conducted using ultrasonic cylindrical guided waves in the laboratory environment. Three different types of mechanical defects - gouge, removed metal, and dent, are fabricated in small diameter, 22.22 mm, aluminum pipes and tested. To efficiently propagate the antisymmetric (flexural) cylindrical guided waves through the aluminum pipe in water, a new transducer holder device is designed. The device uses commercially available ultrasonic transducers that generate compressional ultrasonic waves in the water. The device can change the striking angle of the incident beam from 0° to 51°. With the help of this device the incident angle adjustment and frequency sweeping can be carried out. This is necessary for obtaining the time history of the received signals for various incident angles and signal frequencies; then these time histories are converted to V(f) curves, or received signal amplitude vs. frequency curves. From the amplitude of these V(f) curves the type and extent of the mechanical defects can be estimated. This investigation shows that the new coupler device can be effectively used for health monitoring of underwater pipelines using guided waves.

## INTRODUCTION

It is imperative to regularly check the integrity of pipelines since any collapse and malfunction of pipelines may cost human life. There have been several efforts to efficiently inspect the integrity of pipelines nondestructively. When inspecting pipeline in seawater, there are more challenges due to the marine environment coming from physical, chemical, and biological marine factors, which should be carefully considered in offshore pipeline inspection.

For underwater pipeline inspection visual inspection is very old fashioned but still very useful method. Other methods, such as magnetic, ultrasonic, pressure testing, and in line service inspection, are also used for underwater pipe inspection. These methods are briefly discussed in this paper then ultrasonic guided wave techniques employing ultrasonic guided waves are introduced as a potential method for underwater pipeline inspection.

Three different types of mechanical defects of aluminum pipes are fabricated, and inspected using guided waves. Previously developed sensors and a new transducer holder are introduced for understanding transmitting and receiving mechanisms of guided waves through pipe. With these sensors, experiments are conducted and V(f) curves, of received signal amplitude versus frequency curves, are obtained. Sensitivity of the V(f) curves to the type and the degree of defects is studied.

In summary, the feasibility of cylindrical guided waves for detecting the mechanical defects of aluminum pipes in water is investigated in this paper.

## BACKGROUND

Divers and ROVs (remote operated vehicles) have been used as the principal means of underwater pipeline inspection. The inspection process includes initial visual inspection, then more detailed visual and specialized inspections, and finally, critical inspections employing an appropriate NDT (nondestructive testing) technique. The role of NDT is to discover defects that may threaten the structural integrity of pipelines such as mechanical damage, cracks, and corrosion. Serious defects are sometimes found during annual inspections of areas never before examined by NDT, and the result of these defects may be catastrophic.

The most important NDT is VI (visual inspection), which may be performed by divers and ROVs [1]. Although the human eye's capability of recognizing, interpreting, and analyzing is the most important tool it needs experience. In addition, it cannot be applied for examining internal defects. MPI (magnetic particle inspection), apart from the visual examination, has been the most widely used method under water. It is a very effective method for detecting surface breaking cracks, both under and above water [2]. In addition, the examination has to be performed by divers and areas to be examined have to be cleaned [3]. MFL (magnetic flux leakage) technique has similar properties as MPI and it has been used for in-line service inspection [4,5]. However, MFL technique cannot detect gradual change in the wall thickness, and it cannot be used on pipes made of nonferrous materials. UI (ultrasonic inspection) is another method used for detection of subsurface defects. It is suitable for thickness measurements, weld examinations, and detection of inclusion and internal corrosion [2]. The main purpose of UI has so far been detection and mapping of internal corrosion in risers and pipelines. This UI has also been performed by divers. Other techniques are also available for underwater pipeline inspection such as acoustic emission, radiography, and eddy current [3]. These methods are at different stages of their development.

In addition to UI, the possible NDT methods for detecting the internal corrosion of underwater pipelines may be pressure testing and the use of in-line service inspection devices. Pressure testing is commonly used to verify the integrity of pipelines after installation or repairs regardless of the system's age. The pipeline is filled with water and pressurized, generally to 125 percent of operating pressure, to reveal leaks and flaws [6]. As a routine inspection method, pressure testing has serious limitations as follows: (1) it detects only flaws that are near critical size, and thus gives an operator little or no warning of impending failure; and (2) it requires shutting-in an entire field, and sometimes several fields. Thus, pressure testing is useful for certain purposes but not a broadly applicable means of assurance. The use of in-line service inspection (ILI) devices (more commonly referred to as smart pigs) to measure various characteristics of pipelines has continued to gain acceptance in the pipeline industry [7]. A variety of ILI devices are used to provide information to pipeline operators, such as the types and locations of pipe defects, the radii and locations of bends, and even photographic images. Most of them carry instruments to measure mostly magnetic flux leakage (sometimes ultrasonic signal), which indicate metal loss. ILI devices carry their won batteries, tape recorders, and odometers. Most ILI devices in the United States has taken place in onshore pipeline systems, but the conditions of offshore pipelines are more challenging for using it. Thus, the widespread use of the devices will require substantial advances in technology. In addition, some reports mentioned that smart pig data were often hard to interpret, and in some cases erroneous [6]. Furthermore, to use the smart pigs, pipelines have to allow free passage and must have pig launch and retrieval facilities, which are not always available [8].

Most ultrasonic devices on smart pigs installed for thickness measurements use bulk waves. They use reflection, transmission and scattering of bulk waves by internal defects. However, such use of bulk waves for pipe wall thickness measurement is not very economic because of the inspection time. Thus, many investigators have developed guided wave techniques since these techniques have several advantages over the conventional ultrasonic methods - (1) guided waves can propagate a long distance; and (2) there are several wave modes that can have different degrees of sensitivity to different types of defects. Currently, Guo and Kundu [8,9] experimentally showed that cylindrical guided waves are sensitive to pipe defects. They developed new sensors, in the shape of a conical coupling fluid container and annular shaped couplers, and used inclination angle and frequency sweep technique for their experiments. Many other investigators have tried to inspect for corrosion, cracks, and other types of defects in pipes by cylindrical Lamb waves [10-13]. In these inspections, time histories recorded by the receiver have been carefully analyzed for detecting any small signal reflected by the discontinuity. Although this has been useful for detecting some defects, sometimes such small defect signals may remain undetected. Guo and Kundu [8,9] used V(f) curves to detect pipe defects. However these applications have been limited to inspection of pipelines open to air – not buried underground or immersed in water. So far, it seems difficult to find a literature reporting the application of guided waves to underwater pipeline inspection although some investigators have studied similar problems such as cylindrical guided wave leakage due to liquid loading [10]. Wang [14] reported the applications of guided wave techniques in the petrochemical industry - inspection of pipes, heater tubes, vessels, risers, and heat exchanger tubing.

In this paper, a special transducer holder used for transmitting guided waves is introduced and some mechanical defects in underwater pipe are detected from the change in the V(f) curve. Ideally, the inclination angle and sweeping frequency should be controlled to launch the cylindrical guided mode (or cylindrical Lamb mode) that is most sensitive to the pipe damage and least prone to attenuation. We have designed and fabricated a new transducer holder for efficiently generating cylindrical guided waves in underwater pipes. This new design is described in this paper and some experimental results generated by the new device are also reported.

Figure 1 outlines the evolution of the coupler mechanism. First five designs have been well explained by Guo and Kundu [8,9]. For underwater inspection, the new coupler as shown in Figure 1f is introduced. The designs shown in Figures 1d and 1e are identical except the range of incident angle. For the design in Figure 1d, the maximum incident angle is 27° and it is 51° for the design in Figure 1e. In the design of Figure 1f, the plexiglas is removed from the design of Figure 1e and water is used as a couplant instead of plexiglas and Vaseline. This type of water coupling should be most effective for underwater pipe inspection. The maximum range of transmitter angle is 51° for this case.

## EXPERIMENTAL SETUP

Five different aluminum pipes containing three different types of defects were inspected. The outer and inner diameters of the pipes are 22.22 mm and 19.05 mm, respectively, and the pipes are 610 mm long. The defects are removed metal, dent, and gouge as shown in Figure 2. Most mechanical damage discontinuities are combinations of these primary types of anomalies [15]. Definition of these anomalies are given in Guo and Kundu [8] as follows: (1) dent is a localized depression or deformation in the pipe's cylindrical geometry resulting from an applied force; (2) gouge is a forceful movement of metal in a local area on the surface of a pipe resulting in wall thinning and cold working of the pipe wall; and (3) removed metal is the removal of metal from the pipe surface by an applied force.

Figure 2 shows the defect free pipe and four pipes containing gouge, dent, and removed metal type defects. Two different dimensions of removed metal type defects were used and denoted by the term removed metal (less) and removed metal (more) in Figure 2. The detail description of fabricating process is given by Guo and Kundu [8].

A laboratory made ultrasonic unit was used for generating the signals. Broadband Parametric transducers were excited using a MATEC 310 gated amplifier by tone burst signals from a WAVETEK 395 function generator. Here, the central frequency of the transducers is 500 kHz. The signal after being propagated through the pipe wall was received by a receiver and was amplified by a MATEC 605 gated amplifier. This signal was digitized by the Gage 40 MHz data acquisition board, and then the received signal was analyzed. The computer program computed the average amplitude of the signal in a given time window by continuously changing the frequency of the tone burst signal and then plotted this value against frequency to generate the V(f) curves. Figure 3 shows the experimental setup. Here, the transmitter is held by the newly designed ellipsoidal shaped transmitter holder that is shown in Figure 1f. The receiver was located on the aluminum pipe and wooden supports were used to support the pipe.

## EXPERIMENTAL RESULTS

All experiments were carried out seven times to investigate the consistency of the experimental results. Each test was conducted at different circumferential locations. Figure 4 gives the V(f) curves of the pipe with dent for different incident angles, varying from 0° to 51°. The angle is measured relative to a plane normal to the pipe axis. Figure 4a shows the V(f) curves when the transmitter-receiver arrangement of Figure 1e was used and Figure 4b shows the V(f) curves when the transmitter-receiver arrangement of Figure 1f was used. In Figures 4a and 4b, the strongest signals are obtained for transducer incident angles of 20° and 51°, respectively.

For the experimental setup of Figure 1e, 20° incident angle was selected since it gives the strongest signals and for the experimental setup of Figure 1f, 51° striking angle was selected for the same reason.

Since only one transmitter is used to generate guided waves, non-axisymmetric cylindrical guided waves are generated during experiments. If we use a larger number (four or more) of transducers uniformly placed around the pipe and then excite them in the same phase then we generate strong axisymmetric modes and weak non-axisymmetric modes [9].

Figure 5 shows the V(f) curves for five different pipe specimens that have: (a) no defect, (b) gouge, (c) remove metal less, (d) remove metal more, and (e) dent as shown in Figure 2. For these curves, the transmitter-receiver arrangement is shown in Figure 1e and the transmitter angle is 20°. Multiple curves in each V(f) plot show consistency of experimental results. Although there are some small fluctuations, the overall pattern is consistent. Figure 5f shows the average values of V(f) curves for five specimens. From the figure, it is not very clear that the guided wave technique can detect and distinguish different types and degree of mechanical defects although there are some trends at peak positions.

Figure 6 shows the V(f) curves for the same five pipe specimen studied in Figure 5. However this time, the transmitter-receiver arrangement as shown in Figure 1f is used. The transmitter angle is 51°. As in Figure 5 here also multiple curves in each V(f) plot show the consistency of experimental results. Figure 6f shows the average values of V(f) curves for five specimens. From the figure, it is clear that the guided wave technique can clearly distinguish between different types of mechanical defects. Defect discrimination in Figure 6f is much more than that in Figure 5f.

Figure 7 shows the V(f) curves for the same five pipe specimens. The transmitter-receiver arrangement for this case is shown in Figure 1e and the transmitter angle is 51°. As before, multiple curves in each V(f) plot show the consistency of experimental results. In this case also V(f) curves show sensitivity to defect types. However, for this case, the difference between V(f) curves for gouge and remove metal more is not very clear at the second peak position and the V(f) curves corresponding to these two defects at the first peak position is different from that of Figure 5f.

From these experimental results it is concluded that 51° transmitter angle gives good results for both transmitter-receiver arrangements as shown in Figures 1e and 1f. Between these two the arrangement in Figure 1f appears to be the better arrangement. It should be noted here that the two peaks at 1.26 and 2.02 MHz in the V(f) curves correspond to two cylindrical guided wave modes.

## ANALYTICAL STUDY

To identify the wave modes corresponding to the two peaks in the V(f) curves, the phase velocity dispersion curves are obtained. Cylindrical guided waves are dispersive and the dispersive pattern is complex. For drawing the dispersion curves the material properties of the specimen structure must be known. In this case the specimen is an aluminum pipe surrounded by water or vacuum. When the surrounding medium is vacuum, the pipe walls are traction-free and the computation of dispersion curves is not very difficult. However, the computation becomes more complex when the surrounding medium is water. In addition, many wave modes appear close to one another for aluminum pipes surrounded by water; then it is not easy to identify which modes correspond to the peaks in the V(f) curves.

The material properties of aluminum pipes are as follows: (1) density = 2.7 g/cm$^3$, (2) longitudinal wave speed = 6.32 km/s, and (3) shear wave speed = 3.13 km/s. These values are obtained from material handbook and the attenuations of aluminum pipes are ignored. To plot the peaks of the V(f) curves on the dispersion curves, Snell's law is used as follow:

$$V_{ph} = \frac{V_c}{Sin\theta} \quad (1)$$

where, $V_{ph}$ is the phase velocity of the cylindrical guided wave in the pipe, $V_c$ is the longitudinal wave speed in the coupling medium between the transducer and the pipe (for Plexiglas $V_c$ = 2.77 km/s; for water $V_c$ = 1.48 km/s), and $\theta$ is transmitter angle. From the Snell's law we can calculate the phase velocity of cylindrical guided waves. For plexiglas coupling the calculated phase velocities are 3.56 km/s and 8.10 km/s for 51° and 20° transmitter angles respectively. For water coupling the calculated phase velocity is 1.90 km/sec for 51° transmitter angle. As discussed in the previous section the frequencies corresponding to the two peaks are 1.26 MHz and 2.02 MHz, respectively.

Only 1$^{st}$, 2$^{nd}$, and 3$^{rd}$ flexural wave modes of the pipe in vacuum are shown in Figure 8 since we propagate non-axisymmetric (flexural) modes in the pipe wall with our transmitter arrangement. The 4$^{th}$ and higher order modes are not plotted since those modes are not very strong for our transmitter-receiver arrangement. For comparison purposes longitudinal wave modes are also plotted in Figure 8.

The solid squares of Figure 8 show the experimental points when 51° transmitter angle is used and the solid circles show the experimental points when 20° angle is used. The positions of solid squares are near the asymptote of some flexural wave modes. Since the 51° transmitter angle gives better results than 20° angle for this set of specimen we can conclude that the wave modes near the asymptote of flexural wave modes are more efficient for this inspection. Guo and Kundu [9] and Ghosh et al. [16] observed similar phenomenon for pipes and plates as well. Guo and Kundu [9] used 31° incident angle for their tests. Their peaks are plotted in Figure 8 with solid triangle symbols. These points are also near the asymptote of some flexural wave modes.

Longitudinal and 1$^{st}$ flexural wave modes of the pipe in water are shown in Figure 9. The second and higher flexural modes are not shown because even without those modes too many curves appear in Figure 9.

Two solid squares of Figure 9 show the experimental points when 51° transmitter angle is used. In Figure 9, the corresponding wave modes cannot be clearly identified since there are too many wave modes in the neighborhood of the experimental points. In addition, if the higher order flexural wave modes (2$^{nd}$ and 3$^{rd}$) are plotted then more dispersion curves appear. However, in spite of all those uncertainties one can say that the positions of solid squares are near the asymptote of flexural wave modes. Hence, for modes to be sensitive to defect type the modes must be close to asymptotes of flexural modes.

## SUMMARY AND CONCLUSION

To investigate the feasibility of flexural cylindrical guided waves for inspecting the mechanical defects of underwater pipes, a new coupler mechanism is designed. Incident angle adjustment and frequency sweep technique are used to generate, propagate, and receive the guided waves through the pipes in water. This study shows that the new coupler is an efficient device and the guided wave inspection technique is an effective tool for health monitoring of underwater pipelines. An analytical study is carried out to identify the cylindrical guided wave modes that are sensitive to the mechanical defects. Dispersion curves of aluminum pipe in air and water are shown. Dispersion curves for aluminum pipes surrounded by water show too many modes. From the analytical study, it is shown that flexural cylindrical guided wave modes near asymptotes of the dispersion curves are more sensitive to the defects in comparison to the modes that are not close to the asymptotes.

This study shows the feasibility of using ultrasonic guided waves for underwater pipeline inspection. The new transducer holder and its coupling mechanism have potential to be used for underwater pipe inspection.

## ACKNOWLEDGEMENTS

This research was financially supported by the NSF grants CMS-9896182, CMS-9901221 and the EPRI Grant EP-P241/C110.

## REFERENCES

1. Goldberg, L., 1996, "Diversity in Underwater Inspection", Materials Evaluation, Vol. 54, No. 3, pp. 401-403.
2. Bayliss, M., Short, D., and Bax, M., 1988, *Underwater Inspection*, E. & F. N. Spon, London.
3. Nordbø, H., 1986, "NDE-Overview and Legal Requirements", Submersible Technology, Society of Underwater Technology, Graham & Trotman Ltd., London, pp. 183-187.
4. Upda L., Mandayam, S., Upda, S., Sun, Y., and Lord, W., 1996, "Development in Gas Pipeline Inspection Technology, Material Evaluations, Vol. 54, No. 4, pp. 467-471.
5. Mandal, K., Dufour, D., Krause, T. W., and Atherton, D. L., 1997, "Investigations of Magnetic Flux Leakage and Magnetic Barkhausen Noise Signals from Pipeline Steel", Journal of Physics D: Applied Physics, Vol. 30, No. 6, pp. 962-973.
6. Committee on the Safety of Marine Pipelines, 1994, "Improving the Safety of Marine Pipelines", National Research Council, National Academic Press, Washington, D. C., ISBN 0-309-05047-2.
7. Crump, H. M. and Papenfuss, S., 1991, "Use of Magnetic Flux Leakage Inspection Pigs for Hard Spot Detection and Repair" Materials Performance, Vol. 30, No. 6, pp. 26-28.
8. Guo, D. and Kundu, T., 2000, "A New Sensor for Pipe Inspection by Lamb Waves" Materials Evaluation, Vol. 58, No. 8, pp. 991-994.
9. Guo, D. and Kundu, T., 2001, "A New Transducer Holder Mechanism for Pipe Inspection" Journal of Acoustical Society of America, submitted.
10. Rose, J. L., Cho, Y., and Ditri, J. L., 1994, "Cylindrical Guided Wave Leakage Due to Liquid Loading", Review of Progress in QNDE, Vol. 13A, Ed., D. O. Thompson and D. E. Chimenti, New York, NY, Plenum Press, pp. 259-266.
11. Alleyne, D. and Cawley, P., 1995, "The Long Range Detection of Corrosion in Pipes Using Lamb Waves", Review of Progress in QNDE, Vol. 14B, Ed., D. O. Thompson and D. E. Chimenti, New York, NY, Plenum Press, pp. 2073-2080.
12. Chan, C. W. and Cawley, P., 1995, "Guided Waves for the Detection of Defects in Welds in Plastic Pipes", Review of Progress in QNDE, Vol. 14B, Ed., D. O. Thompson and D. E. Chimenti, New York, NY, Plenum Press, pp. 1537-1544.
13. Cheng, A. and Cheng, A. P., 1999, "Characterization of Layered Cylindrical Structures Using Cylindrical Waves", Review of Progress in QNDE, Vol. 18A, Ed., D. O. Thompson and D. E. Chimenti, New York, NY, Plenum Press, pp. 223-230.
14. Wang W. D., 1999, "Applications of Guided Wave Techniques in the Petrochemical Industry", Review of Progress in QNDE, Vol. 18A, Ed., D. O. Thompson and D. E. Chimenti, New York, NY, Plenum Press, pp. 277-284.
15. Davis, R. J. and Bubenik, T. A., 1996, "The feasibility of Magnetic Flux Leakage In Line Inspection as a Method to Detect and Characterize Mechanical Damage", GRI (Gas Research Institute) Report No. GRI-95/0369.
16. Ghosh, T., Kundu, P., and Karpur, P., 1998, "Efficient Use of Lamb Modes for Detecting Defects in Large Plates", Ultrasonics, Vol. 36, pp. 791-801.

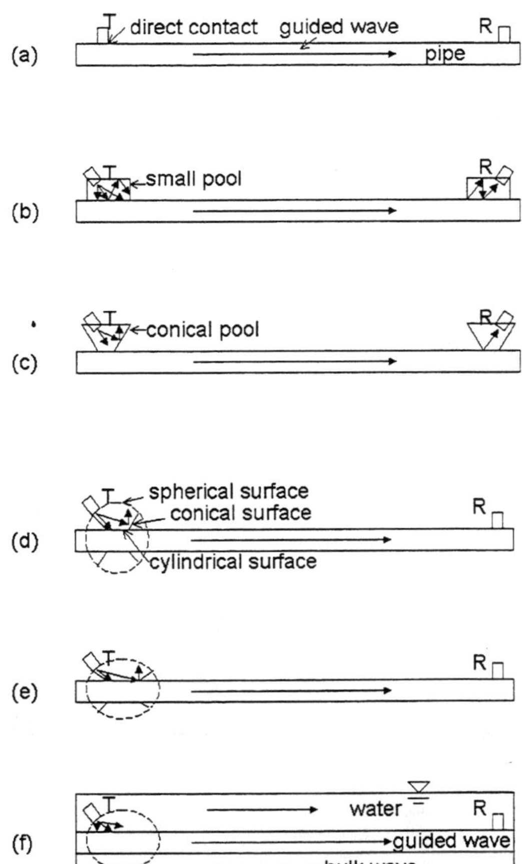


**Figure 1. Evolution of the sensor design: (a) direct contact, (b) small pool of coupling fluid, (c) conical pool, (d-e) annular plexiglas holder, and (f) annular holder without plexiglas.**

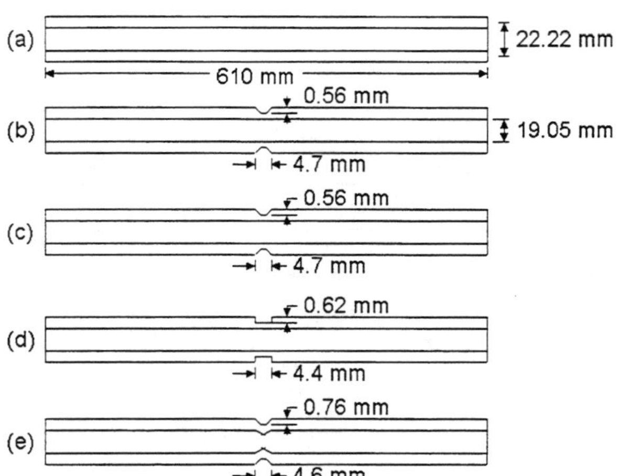


**Figure 2. Aluminum pipe specimen: (a) defect free; (b) gouge; (c) removed metal (less); (d) removed metal (more); and (e) dent. All pipes have same length, outer radius, and inner radius.**

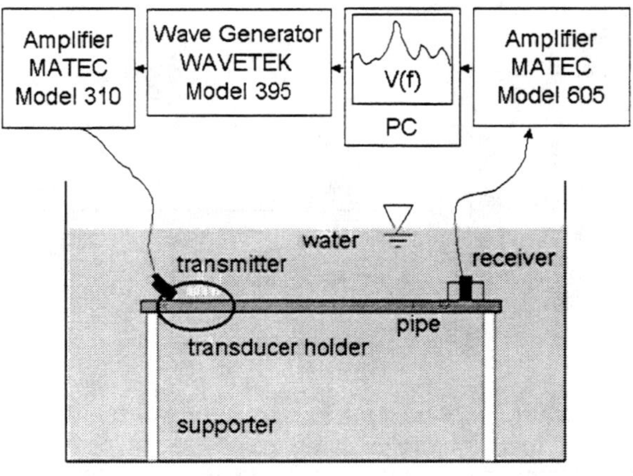


**Figure 3. Schematic of experimental setup.**

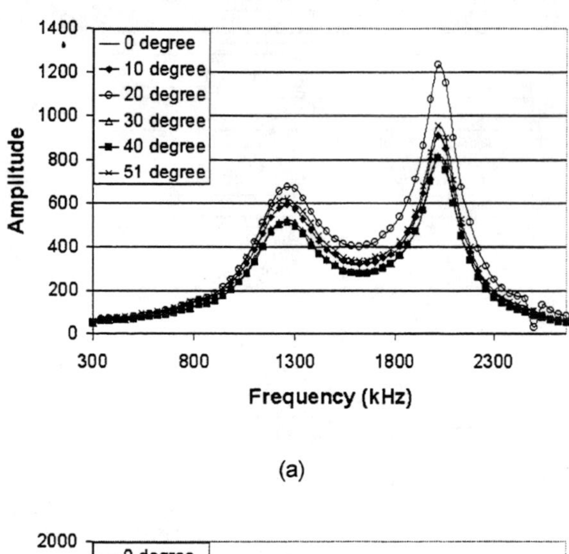


(a)

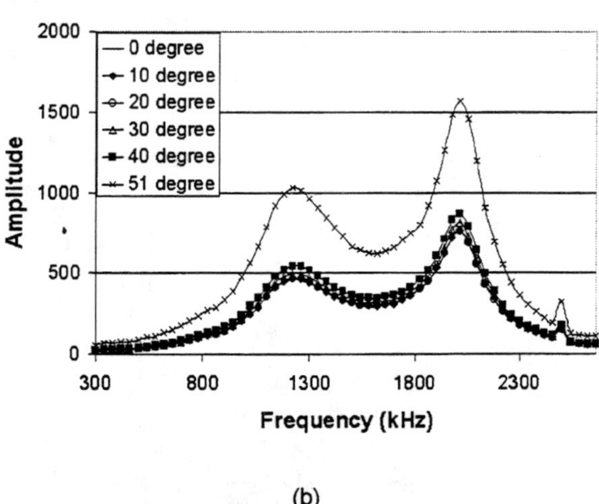


(b)

**Figure 4. V(f) curves of the aluminum pipe with dent for different incident angles using transmitter-receiver arrangement as shown in (a) Figure 1e and (b) Figure 1f.**

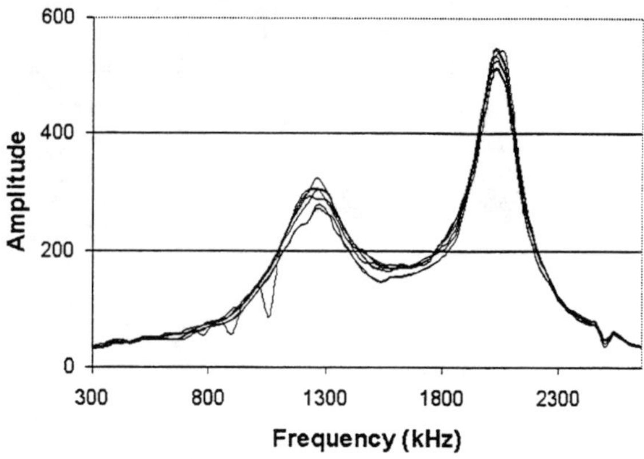


(a)

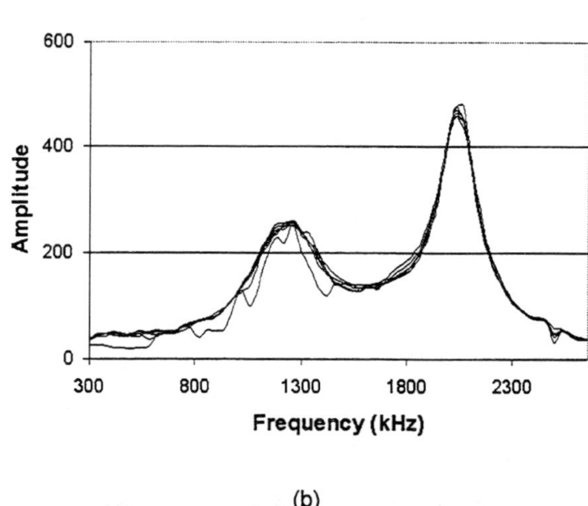


(b)

**Figure 5. V(f) curves of aluminum pipes for 20° transmitter angle for five different defects using transmitter receiver arrangement as shown in Figure 1e: (a) no defect, (b) gouge, (c) removed metal less, (d) removed metal more, (e) dent, and (f) average values.**

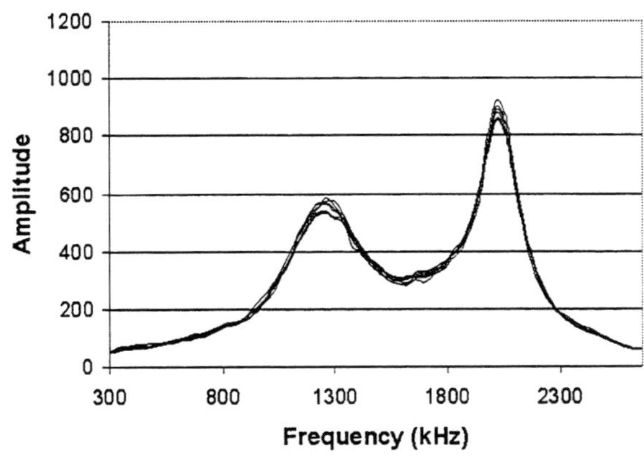


(c)

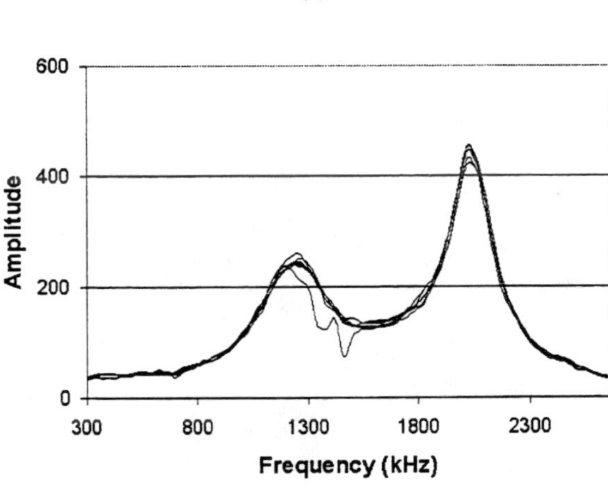


(d)

**Figure 5. Continued.**

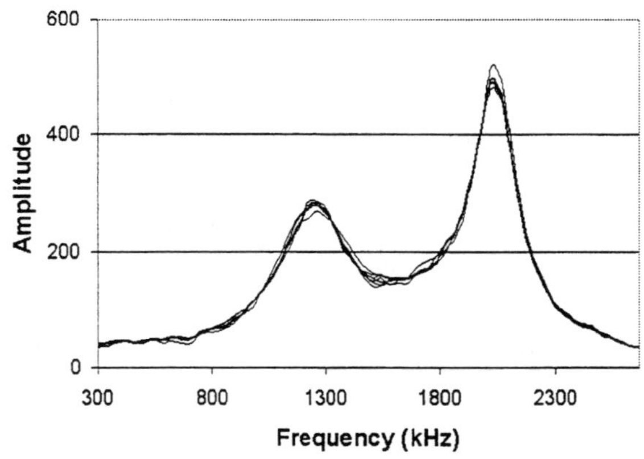


(e)

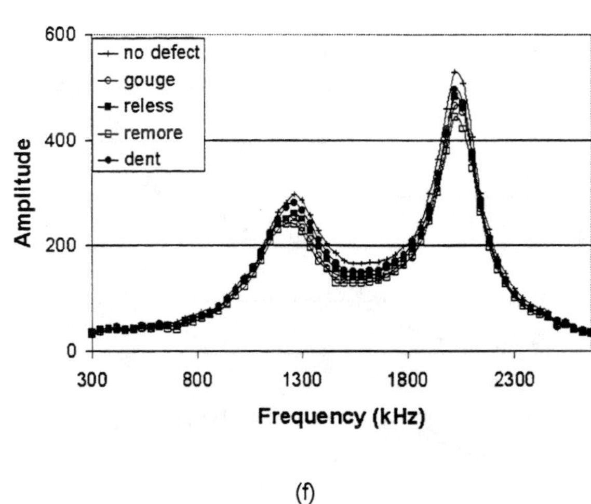


(f)

**Figure 5. Continued.**

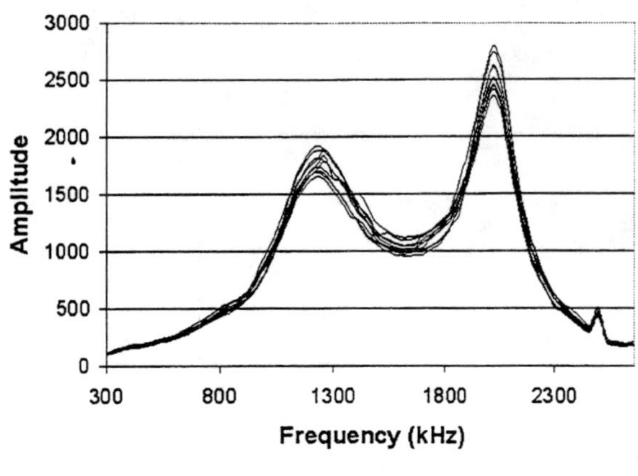


(a)

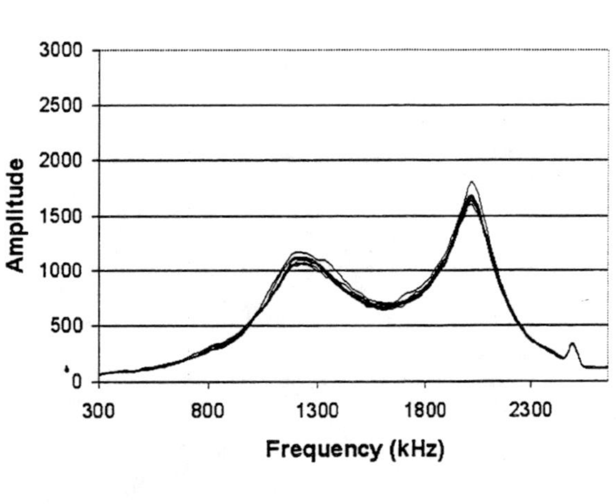


(b)

**Figure 6. V(f) curves of aluminum pipes for 51° transmitter angle for five different defects using transmitter receiver arrangement as shown in Figure 1f: (a) no defect, (b) gouge, (c) removed metal less, (d) removed metal more, (e) dent, and (f) average values.**

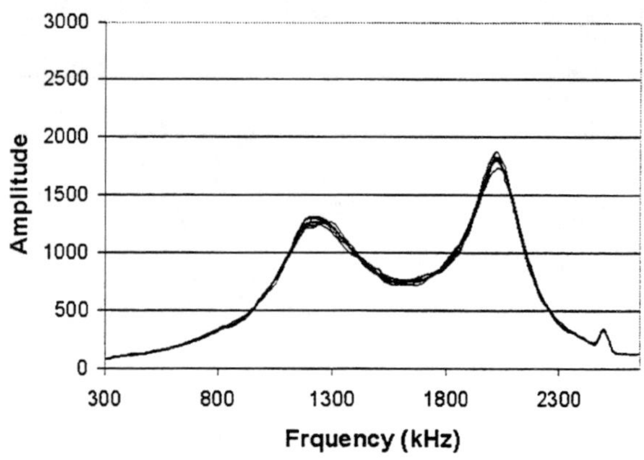


(c)

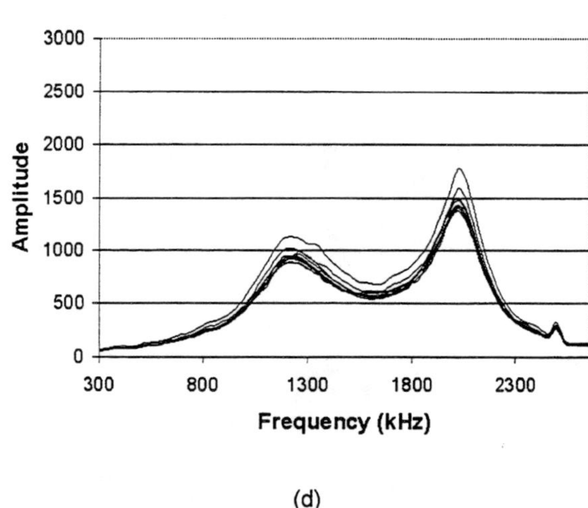


(d)

**Figure 6. Continued.**

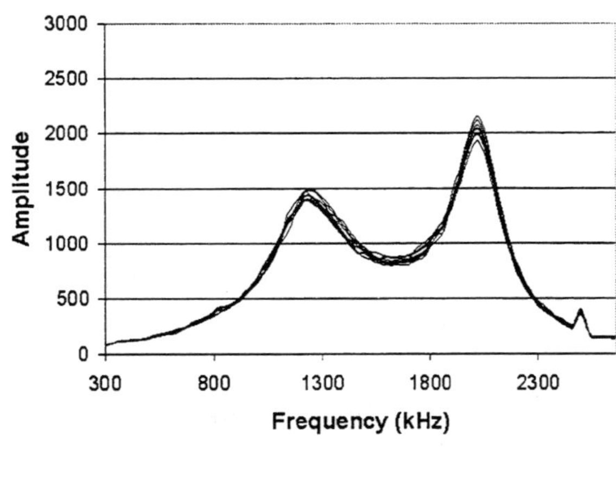


(e)

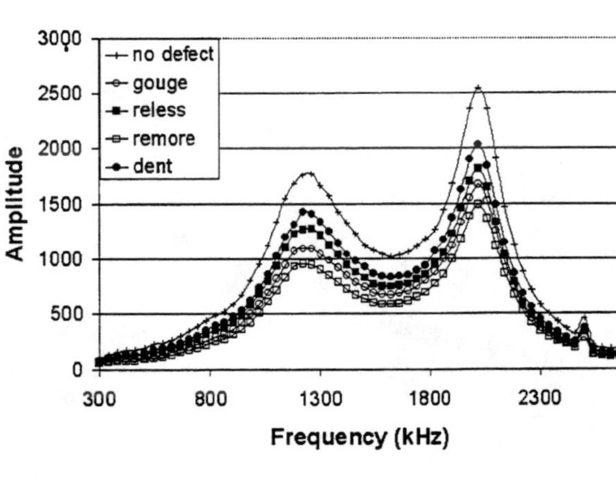


(f)

Figure 6. Continued.

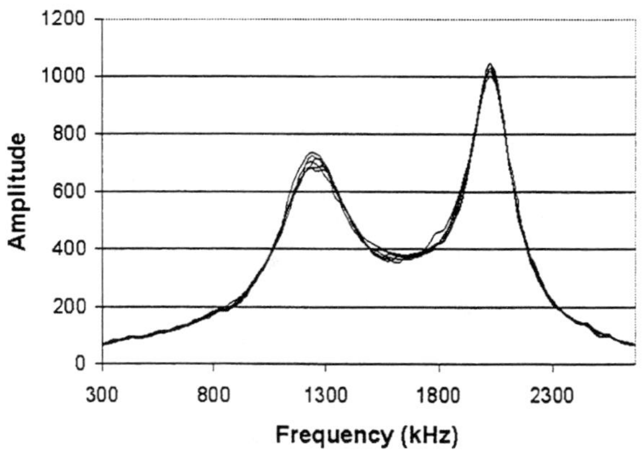


(a)

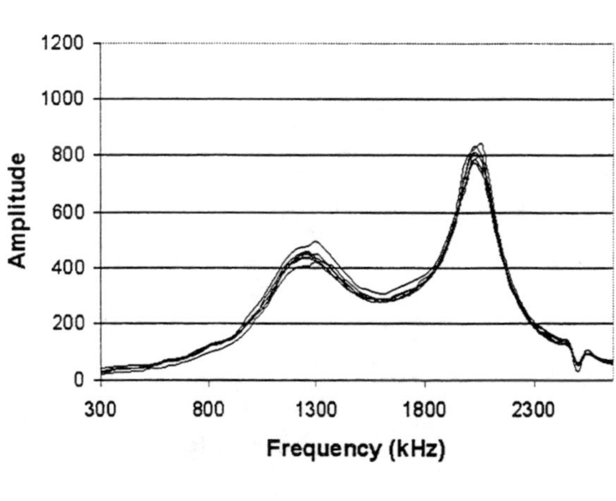


(b)

**Figure 7. V(f) curves of aluminum pipes for 51° transmitter angle for five different defects using transmitter receiver arrangement as shown in Figure 1e: (a) no defect, (b) gouge, (c) removed metal less, (d) removed metal more, (e) dent, and (f) average values. 51° transmitter angle is used.**

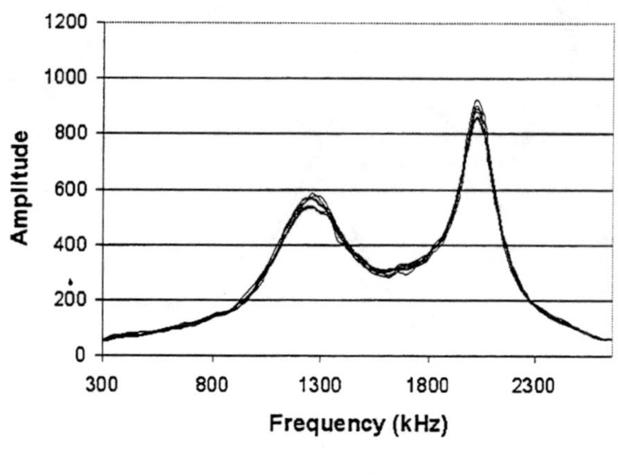


(c)

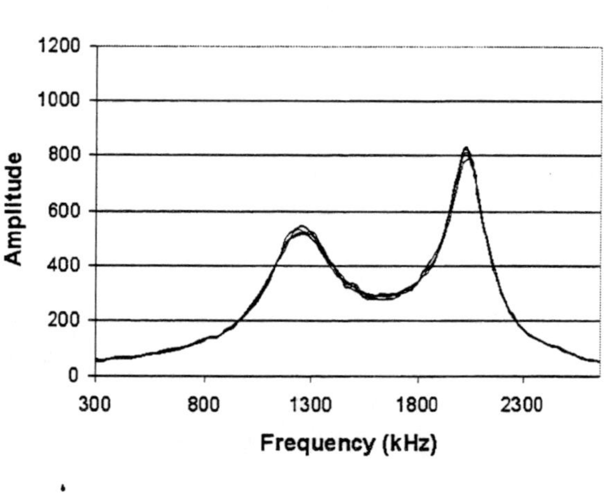


(d)

Figure 7. Continued.

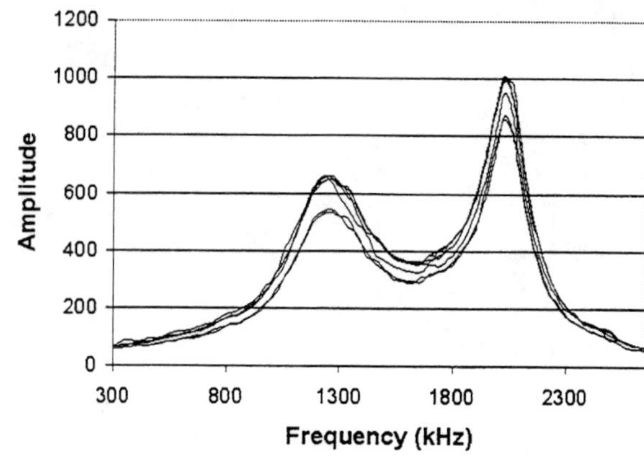


(e)

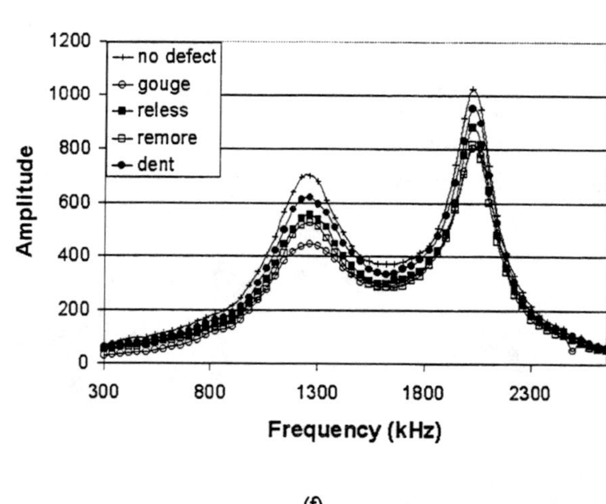


(f)

Figure 7. Continued.

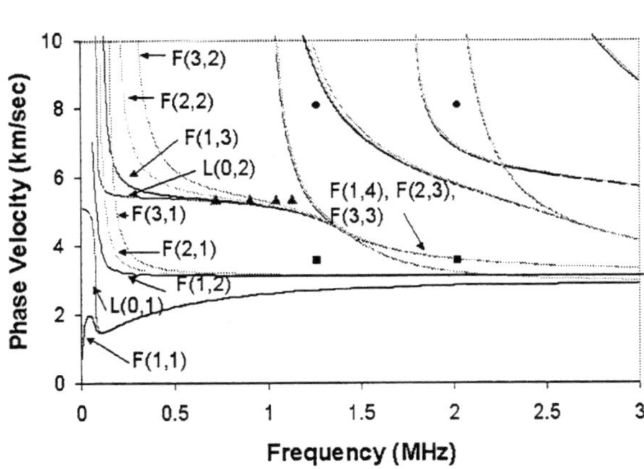


**Figure 8. Phase velocity dispersion curves of the aluminum pipe in vacuum. Longitudinal, 1st, 2nd, and 3rd flexural wave modes are shown. Solid squares and circles correspond to the peak positions of the V(f) curves for 51° and 20° transmitter angles, respectively. Solid triangles correspond to peak positions of the V(f) curves for 31° transmitter angle (Guo and Kundu 2001).**

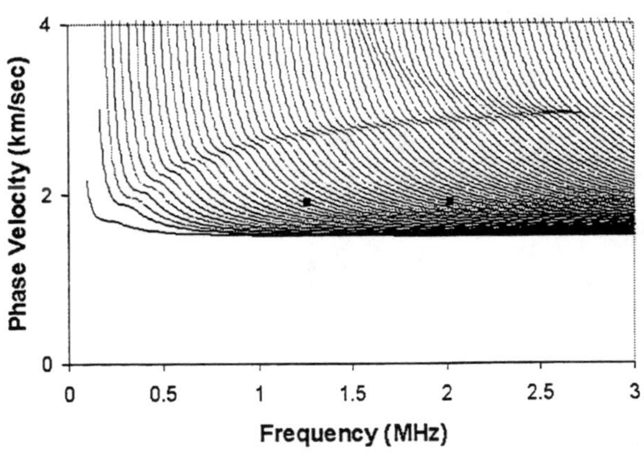


**Figure 9. Phase velocity dispersion curves of the aluminum pipe in water. Longitudinal and 1st flexural wave modes are shown. Solid circles correspond to the peak positions of the V(f) curves for 51° transmitter angle.**

Proceedings of
2001 ASME International Mechanical Engineering Congress and Exposition
November 11–16, 2001, New York, NY
NDE-Vol. 21

**IMECE2001/NDE-25805**

# A MODEL FOR FATIGUE CRACK INITIATION FROM NOTCHES

A. S. Zaki and H. Ghonem

Mechanics of Materials Laboratory
Department of Mechanical Engineering
The University of Rhode Island
Kingston, RI 02881

## ABSTRACT

A model is developed to predict the number of cycles for fatigue crack initiation from a notch. It is based on the concept that the initiation of a fatigue crack occurs when the accumulated plastic shear deformation in the notch root reaches a critical limit which is defined in terms of the threshold stress intensity factor of the material under consideration. A viscoplastic analysis using unified constitutive equations is employed in order to describe the evolution of the notch plastic zone size as well as the stress and plastic strain distributions within this zone on cycle by cycle basis. This analysis takes into consideration the material s time- and cyclic-dependent characteristics. Experimental verification of the model was carried out using specimens made of AM350, an austenitic steel alloy. A series of crack growth measurements were performed in order to calculate the threshold stress intensity factor which is then used to determine the plastic deformation limit of this alloy. The model is used to calculate the number of cycles to crack initiation which is compared to that obtained experimentally under various loading parameters. The correlation between the model prediction and the experimental results are reported and discussed.

## I. INTRODUCTION

Fatigue cracks are known to initiate from stress raisers such as notches, holes and fillets. Predicting life to initiation, i.e. the number of cycles that accumulate before a crack initiates, is very important in applications that cannot withstand having a crack such as aerospace structures. The following is a concise survey of the existing approaches in the literature that are used to determine life to initiation. Crack initiation has been treated in the literature by approaches based on elastic or plastic analysis. Elastic approaches traditionally start with the use of the notch stress concentration factor $K_T$, defined as

$$K_T = \frac{\sigma_{\max}}{\sigma_{\text{nom}}} \tag{1.1}$$

which is the ratio of the maximum stress at the notch root over the nominal stress. This factor is determined from finite element analysis or from closed form analytical solutions. Empirical formula could then be derived relating the $K_t$ for a certain notch radius to the number of cycles to failure [13,15].

This approach, when found inadequate in predicting experimental data, was modified by using the term $\Delta K/\sqrt{\rho}$ instead of $K_T$, where $\Delta K$ is the stress intensity factor of the notch and $\rho$ is the notch radius [2]. The relationship between $\Delta K/\sqrt{\rho}$ and the number of cycles to initiation, $N_i$, can be determined experimentally as a function of the stress ratio R, $R = \sigma_{min}/\sigma_{max}$. Several authors [7,16,2] have shown that there is a threshold value for the $\Delta K/\sqrt{\rho}$ parameter below which fatigue cracks do not initiate.

To account for the plasticity caused by the localized stress at the notch root, other approaches employed the Notch Local Strain criterion. In the former type of analysis, a $K_f$ is introduced as a ratio of the maximum stress in the un-notched specimen to the maximum nominal stress in the notched specimen for the same fatigue life. $K_f$ is related to the stress concentration factor, $K_t$, as follows [13]:

$$K_f = 1 + \frac{K_t - 1}{1 + (\rho'/\rho)^{1/2}} \tag{1.2}$$

where $\rho'$ is a material and stress dependent parameter that defines the distance from the notch tip at which there is no more stress gradient. In this analysis, $K_t$, is expressed as:

$$K_t^2 = K_\sigma K_\varepsilon \tag{1.3}$$

where $K_\sigma = \Delta\sigma / \Delta\sigma_{nom}$ and $K_\varepsilon = \Delta\varepsilon / \Delta\varepsilon_{nom}$ are the elastoplastic stress and strain concentration factors respectively. $\Delta\sigma$ and $\Delta\varepsilon$ are the localized tensile stress and strain ranges at the notch root. Emperical curves are generated correlating $K_f$ and $N_i$ as demonstrated in Ref.[17].

In the Notch Local Strain criterion [15], the local stress and local strain at the notch root are either measured or computed using the finite element method on the basis of the notch geometry and the material cyclic stress-strain curve. Furthermore, in order to consider the combined effect of the stress and strain and operate independent on the notch radius, the quantity $\sqrt{E\Delta\sigma\Delta\varepsilon}$, is introduced as an initiation control parameter, where E Young s modulus, $\Delta\sigma$ is the notch local cyclic stress and $\Delta\varepsilon$ is the notch local cyclic strain. The relationship between $\sqrt{E\Delta\sigma\Delta\varepsilon}$ and $N_i$ for a particular stress ratio is generally defined empirically. However, this approach does not take into consideration the stress or strain gradient ahead of the notch and ignores variations in the stress distribution as the material hardens during the cyclic loading.

For a fully reversed type of loading, $N_i$ can be determined from low cycle fatigue laws, see ref. [6]. Here, the crack initiation is considered equal to the number of cycles to failure, $N_f$, of a smooth specimen subjected to a stress or strain range similar to that experienced by the notch root. In this case, $N_f$ is calculated using Coffin-Manson equation:

$$\frac{\Delta\varepsilon^p}{2} = \varepsilon_f'(2N_f)^c \tag{1.4}$$

where $\Delta\varepsilon^p$ is the notch local plastic strain range, $\varepsilon_f'$ is the fatigue ductility coefficient and c is the fatigue ductility exponent of the material.

Methods employing damage concepts were also developed to calculate $N_i$. Damage accumulation variables were described in terms of the plastic shear deformation [8,9] which assumes that a crack initiates when the plastic shear displacement due to notch bulk plasticity exceeds a threshold value. Hammouda et al. [8,9] described this threshold in terms of the material s mode I threshold stress intensity factor range as well as the cyclic yield stress range.

Other damage based models employed the plastic strain energy approach which acknowledges both the effect of the stress and strain on the fatigue process by considering an energy criterion. Accordingly, a crack is initiated when the summation of the plastic strain energy in the plastic zone around the notch, reaches a critical value. As such, these solutions are not able accommodate the evolution of the hardening characteristics of the notch root material under cyclic loading and therefore, could not provide realistic assessment of the notch crack initiation life.

In order to overcome this deficiency, an attempt is made in this paper to provide a measure of the notch damage taking into consideration the material cyclic viscoplastic response. The next section of the paper presents the basic formulations of a proposed crack initiation model with a focus on a damage criterion and a material model suitable for the use in cyclic loading environment. The initiation model is then applied to determine the number of cycles for the initiation of a crack from a notch and results are compared with those obtained experimentally.

## II. DAMAGE MODEL

The model presented here is based on the concept that damage is measured by the magnitude of plastic deformation in the notch root. Under cyclic loading, the progressive accumulation of the plastic deformation is assumed to occur by a ratchetting process governed by the changes in the kinematic hardening of the notch root material. In the presence of ratchetting, crack initiates when the total plastic shear deformation reaches a critical limit. The determination of the number of cycles to crack initiation would then require the knowledge of two basic items. The first item is a model that is capable of describing the viscoplastic response of the notch root material under cyclic loading. This model would, on the basis of the knowledge of the local stress field in the notch root region, be utilized in a ratchetting procedure to determine the total plastic shear deformation on cycle-by-cycle basis. The second important item required in this work is a crack initiation criterion based on the accumulated plastic shear deformation. Satisfying this criterion by comparing its magnitude and that produced by the ratchetting model under a particular loading conditions, would provide the corresponding crack initiation life. These two requirements will be detailed in the next two sections.

**Damage Criterion:** As a result of the externally applied elastic stress, plastic deformation could occur in the notch root region with a gradient that depends on the applied stress level, notch geometry and material s elastic-plastic properties. Plastic shear deformation, $\Phi_t$, along the length of the maximum shear plane, $\Delta_s$, within the notch plastic zone could be approximated as, see Figure 1, As a result of the stress strain history varies across the plastic zone of the notch root, one could then view $\Phi_t$ as an average quantity estimated across this zone. The average

$$\phi_t = \int_0^{\Delta_s} \gamma_p \, ds \tag{2.1}$$

where $\gamma_p$ is the strain distribution along the plastic zone size

$\Delta_s$. The difficulty in estimating the value of $\Phi_t$ at each loading cycle is the absence of closed form solutions for general strain-hardening material which could be employed to calculate $\gamma_p$ or $\Delta_s$. This could be overcome by the use of a finite element analysis to provide these two quantities on cycle by cycle basis. The success in this approach, however, depends on the availability of a material viscoplastic model that takes into consideration the cyclic evolution of the material hardening characteristics. This model will be detailed in the next section.

Now, if one succeeds in calculating $\Phi_t$ at each loading cycle, an assumption can then be made that the number of cycles to crack initiation is reached when $\Phi_t$ reaches a critical displacement limit, $\Phi_{th}$. The later parameter could be considered a material dependent parameter. One attempt in this regard is to evaluate $\Phi_{th}$ in terms of the material s cyclic ductility limit, which is based on fracture mechanics concepts and assumes that crack initiation from a notch can be described as a function of the crack tip opening displacement. This concept has lead to an expression for $\Phi_{th}$ defined in terms of the threshold stress intensity factor, $\Delta K_{th}$. This definition is written as:

$$\phi_{th} = \frac{1}{\sqrt{2}} \frac{\Delta K_{th}^2}{E \Delta \sigma_{cy}} \tag{2.2}$$

where E is the elastic modulus and $\sigma_{cy}$ is the cyclic yield range. $\Delta K_{th}$ is a material dependent, which is shown to be a function of the stress ratio R as well as test temperature and environment conditions [13,18].

As mentioned above, the number of cycles for crack initiation from a notch is determined by a cycle-by-cycle comparison of equations 2.1 and 2.2. Equation 2.1 requires the knowledge of the strain distribution across the plastic zone of the notch for each loading cycle. In the case of an elasto-plastic material, this can be determined by adopting a constitutive material model which is able to predict the plastic strain flow rate under cyclic loading. Furthermore, the conversion of the plastic shear strain into deformation requires the knowledge of the notch plastic zone size. Equation 2.2 on the other hand would requires the knowledge of $\Delta K_{th}$ as function of the stress ratio R.

## III. MATERIAL MODELING

This section of this paper details the analytical and computational efforts to predict the accumulative plastic strain in the notch. This will be followed by a description of the experimental efforts carried out to determine $\Delta K_{th}$ as a function of the stress ratio, R.

**Material Viscoplastic Model:** In recent years, there has been substantial development in the unified theories of viscoplasticity in which all the aspect of plastic behavior of the material are represented. In unified theories, the prediction of strain rate dependent plastic flaw under monotonic and cycling loading conditions, creep and stress relaxation is obtained from a single set of equations. These equations consists of: (i) a relation between plastic strain and deviatoric stress (flow law), (ii) evolution equations for variables that resist the plastic flow, and (iii) equations relating the plastic strain rate and stress to those variables. These set of differential equations are then solved simultaneously to describe the behavior of the material under different loading conditions.

The viscoplastic equations developed by Chaboche and Rousselier [3,4], were originally applied to superalloys and then extended in the work of Nouilhas [14] to stainless steels. The basic approach in these equations is the use of the yield surface with a center, the motion of which is described by the kinematic variable, X , which represent the back stress effects. On the other hand, the increase in the initial size of this surface, k, is described by the evolution of the isotropic variable, R. The cyclic viscoplastic model can be described as two parts, one related to the viscous flow rule and the second is dealing with the kinematic and isotropic hardening rules.

*Viscous flow rule:* The viscoplastic behavior of the material is linked to the concept of time-dependent viscous stress which is defined as the distance in the deviatoric stress space between the stress state and the projection of the actual elastic domain, i.e.

$$\sigma_v = J(\sigma - X) - R - k \tag{3.1}$$

Where $J(\sigma - X)$ is the Von Mises second invariant defined as

$$J(\sigma - X) = \left[\frac{3}{2}(\sigma' - X') : (\sigma' - X')\right]^{1/2} \tag{3.2}$$

$\sigma'$, $X'$ are the deviatoric components of the applied stress $\sigma$, and the back stress X, respectively. The plastic strain rate as obtained from the normality rule and assuming a condition of viscoplasticity without viscous saturation, is derived as:

$$\dot{\varepsilon}^p = \frac{3}{2} \dot{p} \frac{\sigma' - X'}{J(\sigma - X)} \tag{3.3}$$

The modulus of this rate is written as:

$$\dot{p} = \left\langle \frac{\sigma_v}{K} \right\rangle^n \tag{3.4}$$

n and K are material-dependent parameters that characterize the strain rate sensitivity of the material and $\langle \cdot \rangle$ is the Heaviside function.

*Hardening rules:* The internal changes during each inelastic transient are described by the kinematics hardening terms and is measured by the back stress, X, which could consist of several terms, i.e.

$$X = \sum X_i \tag{3.5}$$

Each of these terms has the general features of expressing the strain hardening as well as the time-dependent recovery. These features are the products of short range and long range internal stresses, see Refs. [3,4,14]. Therefore, when considering a linear summation of theses features in each term of the back stress, $X_i$, could be written as:

$$\dot{X}_i = C_i(a_i\dot{\varepsilon}^p - X_i\dot{p}) \tag{3.6}$$

Where $a_i$ and $C_i$ are material parameters, In addition to the kinematic hardening, flow properties are effected by the isotropic hardening which correspond to the slow evolution of the microstructure associated with cyclic hardening. Contrary to kinematics hardening, the evolution of isotropic hardening is slow. The isotropic hardening, R, which is determined as the difference between the yielding position after a loading cycle and that corresponding to the monotonic loading for the same plastic strain, is written, in absence of a recovery, as [14]:

$$\dot{R} = b(Q - R)\dot{p} \tag{3.7}$$

Where Q is the saturation limit for R at a particular plastic strain and b is a parameter that describes the evolution nature of R. In order to take into consideration the influence of the plastic strain range memorization on isotropic hardening, the work of Nouailhas [14] modified the Q term by adding a slowly diminishing term. However, the memory influence is generally slow and is not considered in this study. In the current model utilization, Zaki and Ghonem [18] considered the evolution of Q, which is the difference between the monotonic and cyclic stress-strain curves, is expressed in terms of its final value $Q_{max}$ and the maximum strain achieved during loading q. This is written as:

$$Q = Q_{\max}(1 - e^{-\mu q}) \tag{3.8}$$

Where, $q = \max(|\varepsilon^p|, q)$. q, for strain controlled cyclic loading, is determined as $q = \Delta\varepsilon^p / 2$. $Q_{max}$, and $\mu$ are determined by fitting the difference between the monotonic and cyclic stress strain curve as presented in ref. [18]. It was also proved that the above model is capable of simulating ratcheting accurately even when the mean stress is negative.

The equations for the material model for load control conditions are summarized as follows:

$$\sigma_v = |\sigma - X| - R - k \tag{3.9}$$

$$\dot{\varepsilon}_p = \left(\left\langle\frac{\sigma_v}{K}\right\rangle\right)^n sign(\sigma - X) \tag{3.10}$$

$$X = X_1 + X_2 \tag{3.11}$$

$$\dot{X}_1 = C_1(a_1\dot{\varepsilon}^p - X_1|\dot{\varepsilon}^p|) \tag{3.12}$$

$$\dot{X}_2 = C_2(a_2\dot{\varepsilon}^p - X_2|\dot{\varepsilon}^p|) \tag{3.13}$$

$$\dot{R} = b(Q - R)|\dot{\varepsilon}^p| \tag{3.14}$$

$$Q = Q_{\max}(1 - e^{-\mu q}) \tag{3.15}$$

$$q = \max(|\dot{\varepsilon}^p|, q) \tag{3.16}$$

$$\dot{\varepsilon}^t = \frac{\dot{\sigma}}{E} + \dot{\varepsilon}^p \tag{3.17}$$

$$\dot{p} = |\dot{\varepsilon}^p| \tag{3.18}$$

**Table 1.** Material parameters for AM355 austenitic steel

| E (GPa) | *k* (MPa) | $a_1$ (MPa) | $C_1$ | $C_2$ | $a_2$ (MPa) |
|---|---|---|---|---|---|
| 195 | 450 | 194 | 2194 | 349 | 639 |
| *K* (MPa) | *n* | $Q_{max}$ (MPa) | *b* | μ | $\sigma_{cy}$(MPa) |
| 280 | 5.6 | 1074 | 19 | 26 | 450 |

The above set of differential and algebraic equations can be solved simultaneously by numerical integration under both strain control and load control conditions. The simulation and experimental stress –strain curve under monotonic loading for a fast strain rate (1E-2 strain/sec) is shown in Figure 2a. Figure 2b. shows the strain-controlled cyclic loading. It is shown that the model simulates the behavior of the material well.

Most metals when subjected to asymmetric stress cycle with a maximum stress exceeding the elastic limit, would produce a progressive plastic strain, i.e. ratcheting. in the direction of the mean stress which gives rise to a small bias in the tensile direction. A typical experimental curve of ratcheting strain versus the number of cycles up to the point of rupture is shown in Figure 3 and Figure 4.

It is shown that ratcheting strain [7] evolves along three different phases. The first phase occupies the first few cycles in which the strain rate is large, this is followed by a secondary "steady state" phase in which the ratcheting rate is approximately constant. The third phase belongs to large strain immediately followed by rupture. Models aiming at predicting ratcheting behavior, generally focus on the first and second phases of ratcheting since the third phase is treacherous due to the fast strain progression to failure [6].

Results of the model prediction of the ratcheting behavior of AM355 steel in terms of plastic strain versus number of cycles at room temperature are shown in Figure 4. Theses results are compared in the same figures with those obtained experimentally for the same load conditions. It is clear that the model formulation, with the modification represented by equations (3.15,3.18), is able to predict ratcheting plastic strain accumulation accurately except for the first cycle where the error is about 15%. The error decreases with the number of cycles. At 1000 cycle the error is 4%.

**Identification of $\Delta K_{th}$:** A series of experiments were performed on flat method was used specimens with center crack to identify the threshold stress intensity factor for different stress ratio R. The shedding to bring the crack growth rate down to less than 1E-5 mm/cycle as shown in Figure 5 for R=0.1.

An Emperical equation taken from [12] was used to fit the relationship between $\Delta K_{th}$ and R as shown in Figure 6. The Emperical equation is as follows

$$\Delta K_{th} = \Delta K_{th}\big|_0 \left( \frac{1+\alpha}{1+\alpha \dfrac{1+R}{1-R}} \right)^m \tag{3.19}$$

where $\alpha \geq 0$ and $m \geq 0$ are to be determined curve fitting. It was found that m=0.5 and $\alpha$=1.

## IV. RESULTS AND DISCUSSION

The configuration of the specimen used to verify the model and approach is shown in Figure (7.a). A quarter of the specimen is modeled with finite element using ABAQUS, in which Chaboche constitutive equations (equations 3.9 to 3.15) were simulated. The stress ratio is always taken to be 0.1 and nominal stress is changed

Using ABAQUS, the plastic zone size was determined as shown in Figure (7.b). The relationship between the applied nominal stress and the notch root stress is shown in Figure (8). It is shown that in the elastic region, the stress at the notch root is about three times the nominal stress as predicted from the elastic stress concentration charts [13]. As the material start yielding the ratio between the stress at the root to the mean stress is decreased. Moreover, by cycling the loading, the strain accumulation in the plastic zone is shown in Figure (9). It is shown that the rate of plastic strain accumulation decreases dramatically after the first few cycles, with is confirmed from the experimental testing shown in Figure (4).

It is also shown in Figure (4) that after the first few cycles, the plastic strain accumulation becomes linear with the logarithm of the number of cycles. This approximate relation is used to calculate the number of cycles to initiation based analytically.

From the pervious analysis, It is shown that the technique relies heavily on accurate modeling the behavior of the material. For the given notch radius, and a stress ratio of 0.1, the prediction of crack initiation for different maximum stress at the notch root is shown in Figure (10).

The number of cycles to initiation was determined experimentally by cycling a number of specimens for a stress ratio of 0.1 and different nominal stress and using potential drop to detect the crack initiation. The number of cycles to initiation determined experimentally is shown in Figure (10). It is shown that the crack initiation criteria under-estimates the number of cycles to failure slightly. This is attributed to the dependency of the initiation criteria on $\Delta K_{th}$ which is conservative. The model will be tested with other initiation criteria.

## V. CONCLUSIONS

A concept of fatigue crack initiation in a notched specimen subjected to cyclic loading is developed. The concept is based on strain accumulation in the plastic zone due to ratchetting. The initiation model is dependent on the material ratchetting behavior, the size of the plastic zone around the notch which depends on the geometry of the notch, the maximum stress and the material, the material stress intensity factor.

The significance of the model is that it makes use of a viscoplastic material model to simulate the hardening behavior of the material, predict the plastic zone size and estimate the plastic strain accumulation. Consequently, the crack initiation model is dependent on having an accurate material model. Experimental results agree with simulation but more validation for different notch radii is in progress.

## REFERENCES


[1] Bathias, C., Gobra, M. and Aliaga, D., "Low-Cycle Fatigue Damage Accumulation of Aluminum Alloys", Low Cycle Fatigue and Life Prediction, ASTM STP-770, ASTM, Philadelphia, PA, pp. 23-44, 1982.

[2] Boukharouba, T., Tamine, T., Niu, L., Chehimi, C., and Pluvinage G., "The Use of Notch Stress Intensity Factor as a Fatigue Crack Initiation Parameter", Engineering Fracture Mechanics, Vol. 52, No. 3, pp. 503-512, 1995.

[3] Chaboche, J.L., and Rousselier, G., "On the Plastic and Viscoplastic Constitutive Equations-Part I: Ruled Developed with Internal Variable Concept," Transactions of the ASME, Journal of Pressure Vessel Technology, Vol. 105, pp. 153-158, 1983.

[4] Chaboche, J.L., and Rousselier, G., "On the Plastic and Viscoplastic Constitutive Equations-Part II: Application of Internal Variable Concepts to the 316 Stainless Steel," Transaction of ASME, Journal of Pressure Vessel Technology, Vol. 105, pp. 159-164, 1983.

[5] Chow, C. L. and Wei, Y., "A Damage Mechanics Model of Fatigue Crack Initiation in Notched Plates", Theoretical and Applied Fracture Mechanics, Vol. 16, pp. 123-133, 1991.
[6] Coffin L. F. "The deformation and fracture of a ductile material under super imposed cyclic and monotonic strain",Achievement of High fatigue Resistance in Metals and Alloys, ASTM, STP-467,53-76, 1970.
[7] Ferreira, J.M. and Costa, J.D., "Fatigue Crack Initiation in Notched Specimens of 17Mn4 Steel", International Journal of Fatigue, Vol. 15, No. 6, pp. 501-507, 1993.
[8] Hammouda, M. M., Smith, R. A. and Miller, K. J., "Elastic-Plastic Fracture Mechanics for Initiation and Propagation of Notch Fracture Cracks", Fatigue of Engineering Materials and Structures, Vol. 2, pp. 139-154, 1979.
[9] Hammouda, M. M. and Miller K. J., "Prediction of Fatigue Lifetime of Notched Members", Fatigue of Engineering Materials and Structures, Vol. 2, pp. 377-386, 1980.
[10] Hibbit, Karlsson & Sorenson, Inc., "ABAQUS Finite Element Analysis Program ver. 6.0", 1998.
[11] Iino, Y., "Local Fatigue Damage Accumulation around Notch Attending Crack Initiation", Metallurgical and Materials Transactions, A, Vol. 26A, pp. 1419-1430, 1995.
[12] Kujawski, D. and Ellyin F., "A Unified Approach to Mean Stress Effect on Fatigue Threshold Conditions", International Journal of Fatigue, Vol. 17, No. 2, pp. 101-106, 1995.
[13] Neuber, H., "Theory of Stress Concentration for Shear-Strained Prismatical Bodies with Aribitrary Nonlinear Stress-Strain Law", Transaction of the ASME, Journal of Applied Mechanics, pp. 544-550, 1961.
[14] Nouailhas, D., "Unified Modeling of Cyclic Viscoplasticity: Application to Austenitic Stainless Steels," International Journal of Plasticity, Vol. 5, pp. 501-520, 1989.
[15] Saanouni, K. and Bathias, C., "Study of Fatigue Crack Initiation in the Vicinity of Notches", Engineering Fracture Mechanics, Vol. 16, No. 5, pp. 695-706, 1982.
[16] Smith, R. A. and Miller, K. J., "Fatigue Cracks at Notches", International Journal of Mechanical Science, Vol. 19, pp. 11-22, 1977.
[17] Zheng, X.-L, "Modelling Fatigue Crack Initiation Life", International Journal of Fatigue, Vol. 15, No. 6, pp. 461-466, 1993.
[18] Zaki, A. S., and Ghonem H., "Modeling the Ratcheting Phenomenon in an austentic steel at room temperature", Proceeding of the ASME 2000 Design, stress analysis and failure prevention conference, Sept. 10-13, 2000, Baltimore, Maryland, USA.

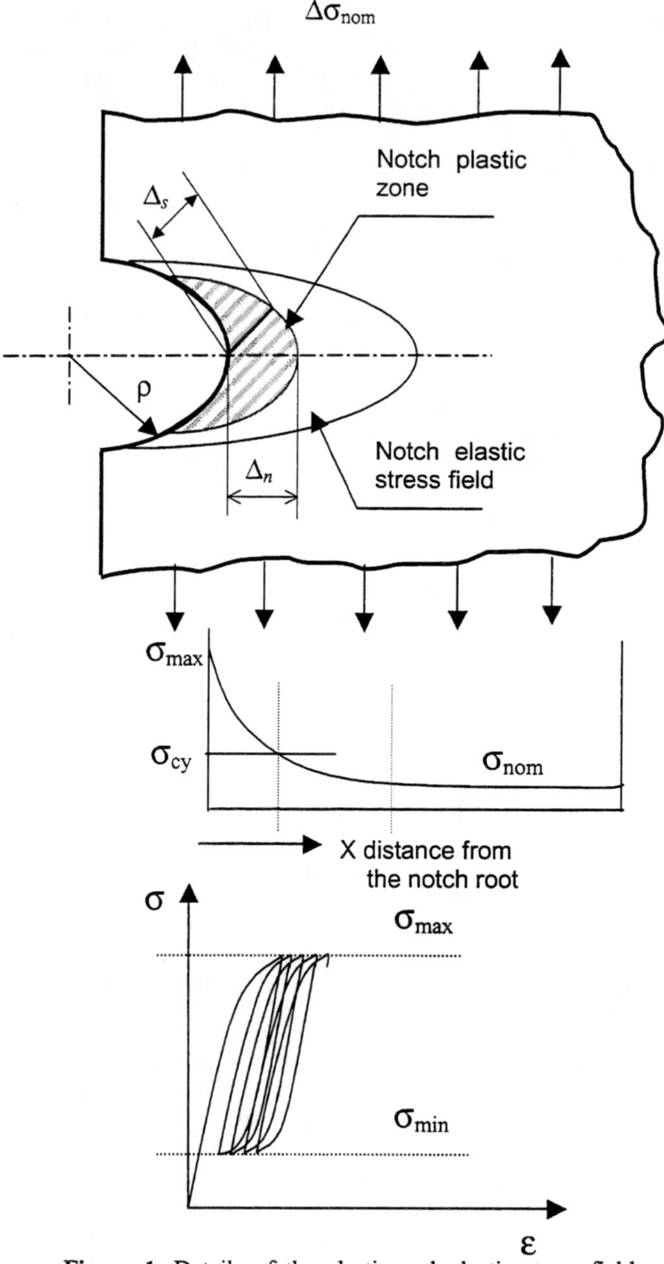


**Figure 1.** Details of the elastic and plastic stress fields at a notch
(a) Schematic of the notch and stresses
(b) Stress field vs. the distance from the notch root
(c) Strain accumulation at the notch root due to ratchetting

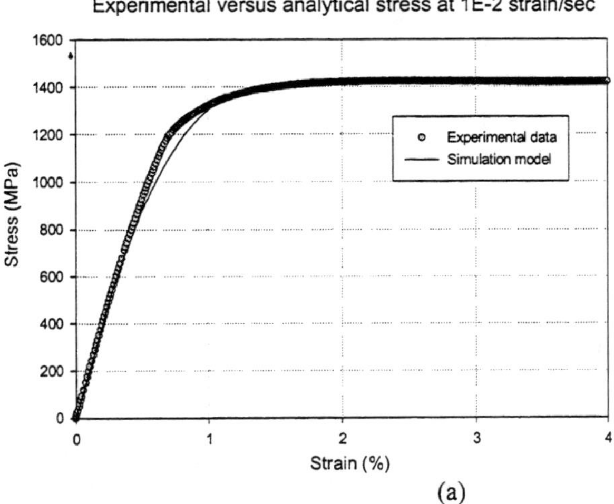


(a)

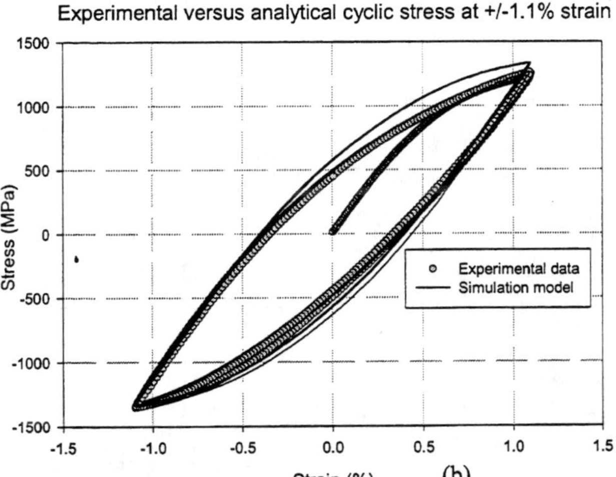


(b)

**Figure 2.** Experimental data versus simulation model
(a) Monotonic loading at 1E-2 strain/sec
(b) Strain controlled cyclic loading (8 cycles)

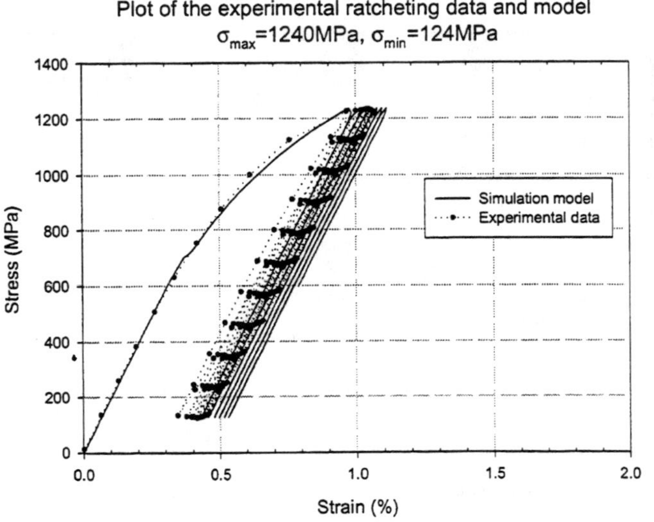


**Figure 3.** Experimental versus simulation results for ratcheting (10 cycles)

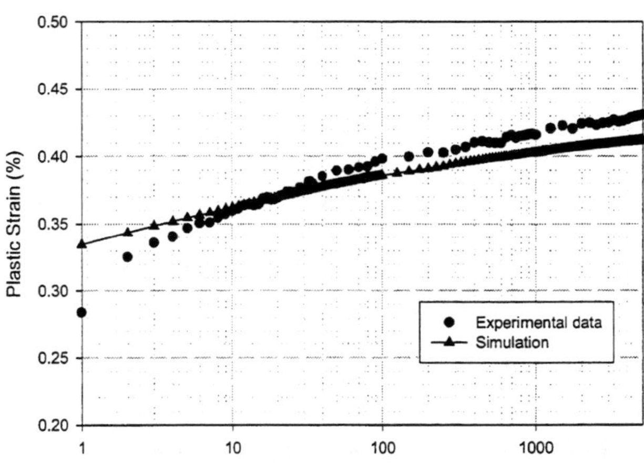


**Figure 4.** Experimental vs. simulation results for plastic strain accumulation with the number of cycles

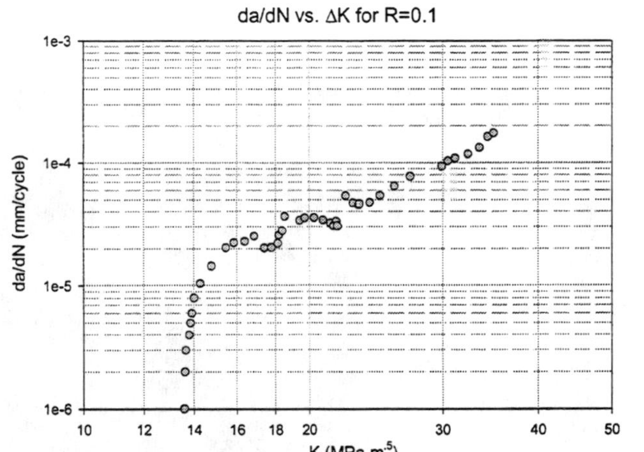


**Figure 5.** Fatigue crack growth behavior for a stress ratio of 0.1. .$\Delta K_{th}$ is taken for da/dN=$10^{-7}$ mm/cycle

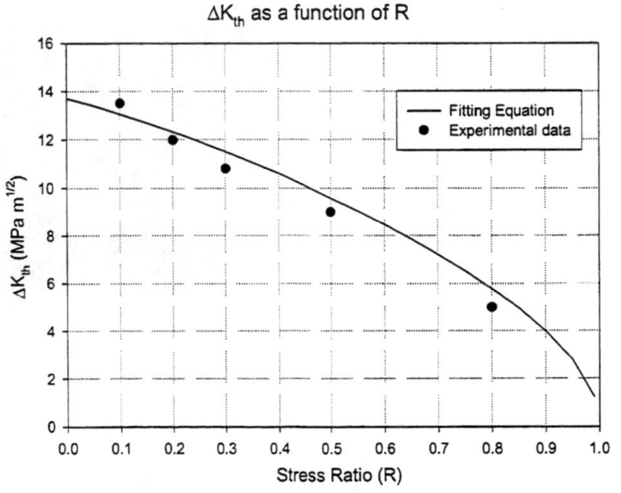


**Figure 6.** $\Delta K_{th}$ as a function of stress ratio, R

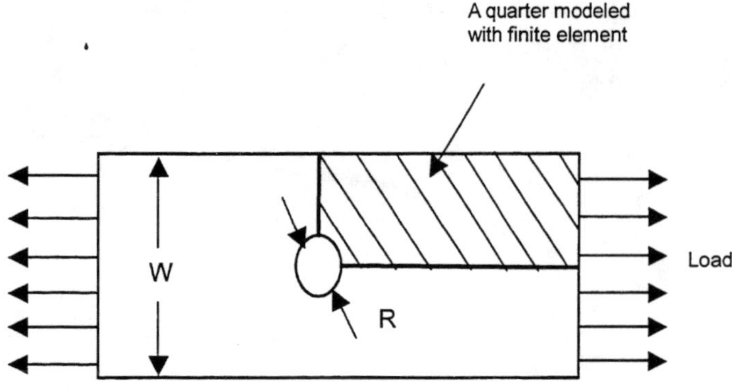


**Figure 7.** (a) Schematic of the specimen dimensions used for finite element simulation and testing, R=0.794 mm, W=7.5mm.

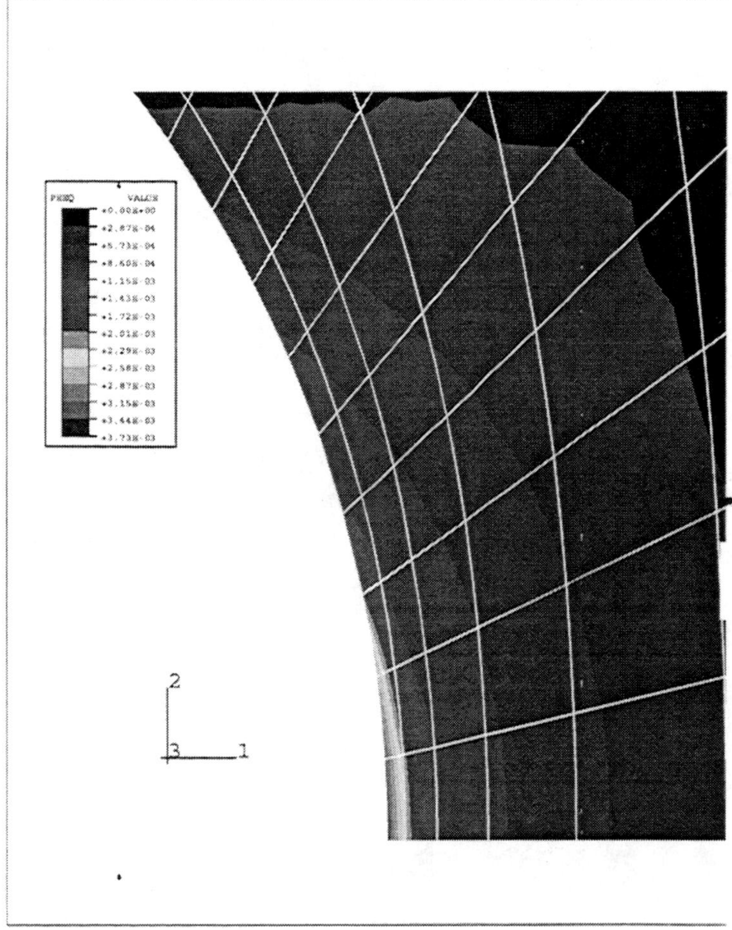


**Figure 7.** (b) Finite element analysis of a quarter of the specimen showing the plastic zone size and the plastic strain distribution at the second cycle.

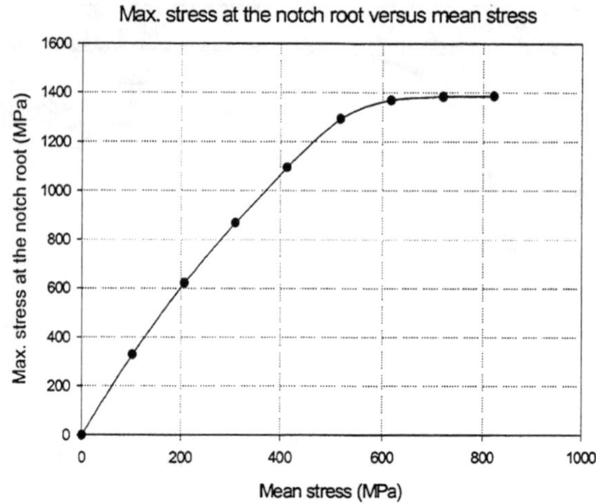


**Figure 8.** FE estimation of the maximum stress at the notch root versus the mean stress outside the notch.

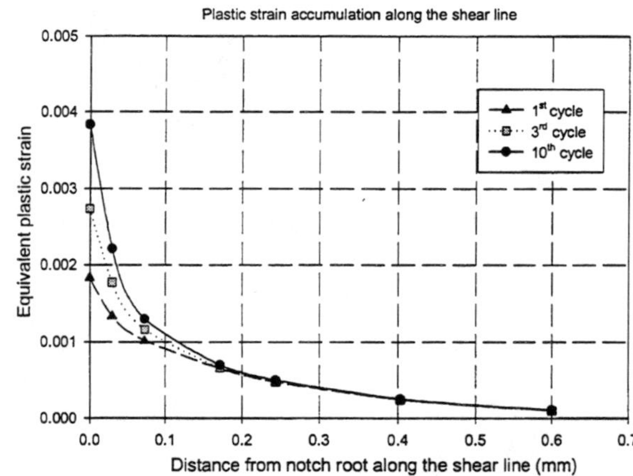


**Figure 9.** FE simulation showing the plastic strain accumulation along the shear line as a function of the distance from the notch root.



**Figure 10.** Number of cycles to initiation as a function of the maximum stress at the notch root for a stress ratio of 0.1.

Proceedings of
2001 ASME International Mechanical Engineering Congress and Exposition
November 11–16, 2001, New York, NY
NDE-Vol. 21

IMECE2001/NDE-25806

# Atomic Force and Ultrasonic Force Microscopic Surface Characterization of Laser Treated Head Sliders

**Carl Druffner, Edward J. Schumaker, Liming Shen, and Shamachary Sathish**
Center for Materials Diagnostics, University of Dayton, 300 College Park, Dayton, Ohio 45469-0121
USA

**Ganesh N. Raikar and Amarjit S. Brar**
Seagate Technology, 7801 Computer Avenue South, Bloomington, Minnesota 55435-5489 USA

## ABSTRACT

The surface of laser treated hard disc drive head sliders has been imaged using the Atomic Force Microscope and Ultrasonic Force Microscope. The contrast of the surface topography image from the Atomic Force Microscope is compared with the elasticity image generated by the Ultrasonic Force Microscope on the same sample region. Images of microcracking in the laser treated regions are presented. The possible reasons for the development of microcracking and the enhanced contrast of the Ultrasonic Force Microscope imaging of these microcracks are discussed.

## INTRODUCTION

In a computer Hard Disc Drive (HDD), the read/write chip, which flies over the magnetic disc storage surface, is called a head slider. As the areal densities of HDD increase and the flying heights decrease, components such as the head slider now measure a mere area of 1mm × 1mm. Being able to obtain high-resolution measurements of the surface structures of such micro-devices is of increasing importance for HDD industry. Currently laser micro-processing is being used to make modifications to several of the hard drive components during their fabrication. Laser micro-processing of these components is being preformed to address application issues ranging from producing specific surface texturing in order to reduce the effects of stiction, to surface cleaning to remove submicron debris or organic films, and to adjust the surface crown to improve the slider flying characteristics. [1]

In order to produce magnetic storage densities exceeding 1GB/in$^2$, a super smooth finish is required on the hard disk surface. Such a smooth surface finish can lead to unacceptable stiction between the head slider and the hard disk surface when the drive is turned off and then on. To reduce the stiction problems, a precise pattern of bumps (landing zone texturing) is usually done on the hard drive but texturing of the head slider is also possible. The typical disk surface texture design is a bump pattern with the bump size of 15-25nm high and having diameters of 10-15μm [2]. Surface texturing reduces the contact area between the hard disc and the head slider - instead of having two smooth surfaces coming into contact with each other - so that the head slider comes to rest on the bumps. Various techniques such as mechanical abrasion, chemical, or laser treating are in use for landing zone texturing of a hard disk surface [3]. Another key element for increasing disc storage density is to minimize the head slider's flying height above the disk surface to about 20nm or so. The crown of its air bearing surface determines the slider's flying height above the hard disk. The head slider crown may need minute adjustments in order for it to have a stable flying height and avoid possibly bumping the disk.

The treatment of a head slider is significantly different than for the landing zone texturing of hard disk surface because the head sliders are a composite made of ceramic material, $Al_2O_3$ - TiC with a magnetic Permalloy read/write coil embedded in an amorphous alumina overcoat. While the hard disk is made from a uniform single phase material, the head slider is a multiphase material with comparable hardness. High hardness, high modulus and optimum toughness of the ceramic can make it difficult and hard to process by mechanical methods especially while working with a device surface area on the order of a square millimeter. Recently, laser cleaning, ion beam etching, as well as, laser curvature adjustment technique (LCAT) have been shown to be very promising methods for modifying head slider surfaces, [1,4-6]. Wahl and Talk have numerically simulated slider surface texturing to show that stiction could be reduced with minimal impact on flying behavior [7].

When attempting to make minute changes to a device as small as the head slider, laser treatment offers several advantages. The major advantages of laser tooling is it is non-contact, no necessity of etching chemicals, abrasion less, eliminates tool wear, tool deflections, vibrations, cutting force and reduces the limitations on structure formation [8]. However, laser cutting by producing high temperature gradients within the material may cause stress cracking. Thus it is important to characterize the laser treated head sliders in order to optimize the tooling conditions and to produce thermally induced stress and crack free head sliders.

The multiphase ceramic composite materials used for head sliders have an average grain size of the order of one micron. Observation of microcracks in such nano-composite materials is a difficult task. Standard optical microscopes are fundamentally limited by optical diffraction to 0.2 μm resolution. Although acoustic microscopy has been used efficiently for observation and detection of cracks, its resolution is limited by the sound wavelength. Scanning Electron Microscope (SEM) can be used but extensive sample preparation is needed and the SEM does not produce a 3D image. One of the most powerful tools that is routinely used for characterization of nanostructured materials is the Atomic Force Microscope (AFM).

The Atomic Force Microscope provides near atomic resolution on both conducting and non-conducting materials in a wide variety of media [9,10,11]. Conventional AFM and modified AFMs have been extensively used for characterization of surface topography, friction and wear properties of head sliders and hard disk surfaces [2,5,12-17]. While most of these studies have been performed on neat head slider materials, very few studies have been reported on laser treated materials. In a recent study on magnetic thin film heads of alumina Chekanov et al. has reported observation of microcracks generated due to thermal fatigue [15-16].

In this paper we present an investigation of ceramic composite head sliders, which are constructed mainly of $Al_2O_3$-TiC, using AFM and Ultrasonic Force Microscopy [UFM]. The $Al_2O_3$-TiC ceramic composition is in the ratio of approximately 70% $Al_2O_3$ and 30% TiC. While AFM has been used to obtain the surface topography map, the UFM has been utilized to gain information about elastic property variation at nanoscales. AFM and UFM images have been obtained on untreated as well as laser treated regions of the head sliders for comparison.

## MATERIALS AND METHODS

A Digital Instruments Dimension 3000 Nanoscope IIIa AFM has been used to characterize the surface topography. To examine the elastic properties of the surface, the AFM was modified to measure the surface displacements generated by propagating an acoustic wave through the sample. There are several different methods of combining ultrasonics with an AFM [18-21]. For this work, the AFM was operated in conventional contact mode with a piezo-electric transducer bonded to one face of the sample. The transducer was excited by a low amplitude continuous wave radio frequency signal in the range of 400 to 1000 kHz. The opposite side of the sample faces an AFM tip, which is being utilized to detect and map the surface displacements generated by the acoustic wave. At low amplitude excitations, it has been shown by Burnham *et al.* that the amplitude detected by the AFM tip is proportional to the elastic stiffness of the material under the tip [18].

In this study, silicon nitride AFM tips were used for the surface topography and elastic modulus mapping. The cantilever was V shaped with a length of 193 μm and a width of 205 μm and has a manufacturer quoted spring constant of 0.12 N/m. Generally the tip has a pyramidal shape with a square base and the nominal radius of the tip curvature is 20-60 nm.

The dimensions of the head sliders were 1.3 mm x 1 mm with a thickness of 0.3 mm. The imaging area during scanning varied from 25 μm × 25 μm down to 500 nm × 500 nm with the scanning speed usually between 0.5 and 1 Hz. The ultrasonic images were captured at frequencies ranging from 450 kHz up to 905 kHz. The head sliders were ultrasonically cleaned in deionized water and dried with acetone before being mounted on the transducer for imaging. A tiny drop of honey was used as the medium between the transducer and the head slider to couple the acoustic waves to the sample.

## RESULTS

Surface topography and ultrasonic images of the head sliders were obtained on laser treated and untreated regions of the head sliders (slider #1, #2 and #3). The AFM and UFM images of the same region were acquired simultaneously. The simultaneous image acquisition was done by using the electronic feedback signal from the AFM's topography image and feeding it through a filtering and separating electronic loop in order to separate the ultrasonic image (high frequency) from the surface topography (low frequency). A Stanford Research

Systems lock-in amplifier is then used to measure the magnitude and phase of the detected ultrasonic signal with respect to the input signal from the Hewlett Packard function generator. The amplitude or phase data was sent to the computer and correlated with the scanning point to generate the ultrasonic image.

While acquiring AFM images was straight forward, for obtaining the UFM images the amplitude and frequency of the acoustic waves had to be optimized. The optimization was performed by two different methods. In the first method the amplitude and the frequency of the input signal were manually varied and the output signal of the lock-in-amplifier was monitored for the maximum stable transmission signal. Too high of a power input into the transducer could bounce the AFM tip off the sample surface resulting in an unstable signal and blurred images. Images were acquired at the frequency with the highest amplitude response while still maintaining a stable signal.

In the second approach, a Hewlett Packard Network Analyzer (Model #8753D) was employed. The transducer was connected to the input side and the output from the AFM tip was connected to the output end of the network analyzer. The analyzer treats the entire system as a two port network. The $S_{21}$, which is defined as the output energy/input energy, was monitored as a function of input frequency. The maximum transmission occurs at a specific frequency for a given configuration. The best transmission frequency was noted and used to set the function generator conditions. Although the second method was quite fast, it gave the same conditions that were obtained with manual tuning method.

## LASER TREATED REGION

Figure 1 shows an optical image of the laser treated region of the slider #1. The laser lines and the position of the cantilever can be seen in this image.

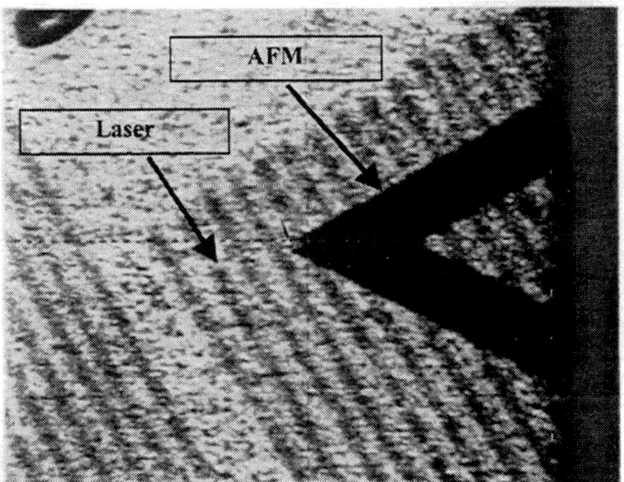


**Figure 1: Optical Image of a laser shaped surface**

Figure 2 shows both the AFM surface topography image and the UFM amplitude image obtained on slider #1. The maximum height in the image is about 1 micron. Several regions with this height are observed in the image. Faint features indicating presence of a crack are also observed (indicated by the arrows), but the cracks are buried under surface topography information. Attempts to improve the contrast of the image by online real time image enhancement procedures did not provide much clearer evidence of the presence of the crack. The UFM image of Fig. 2 is of the same region and was acquired at an excitation frequency of 508 kHz. The contrast in the UFM image is much better than the AFM image and several cracks can be clearly identified.

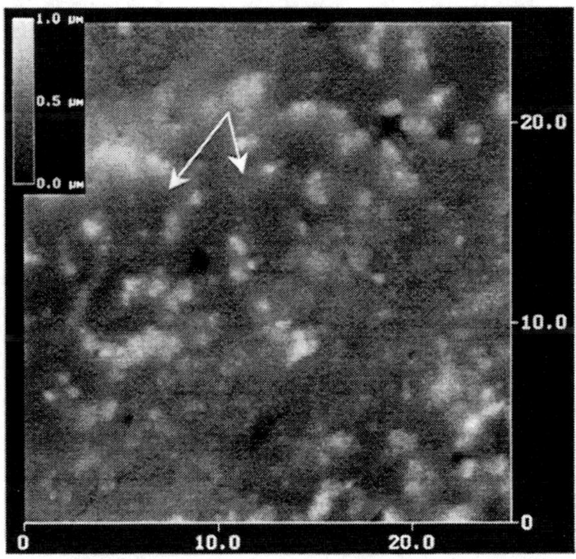

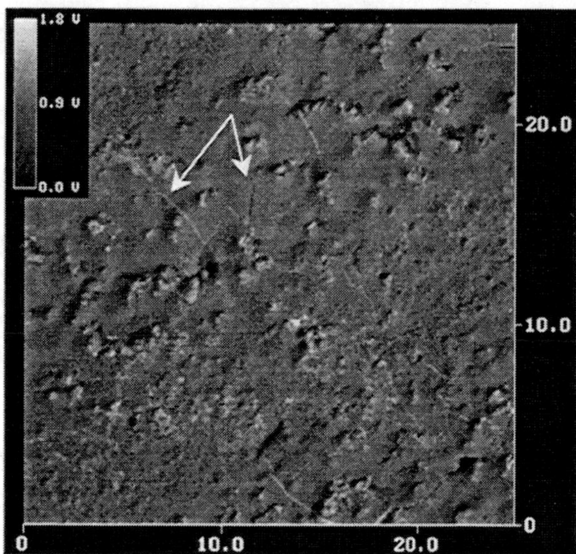


**Figure 2: a) On the top is an AFM topography image. b) On the bottom is a UFM amplitude image at the excitation frequency of 508 kHz. Scan area is 25 μm × 25 μm on slider #1**

A surface topography image of the magnified area of the crack region is shown in Fig. 3a. Even at higher magnification the crack features are not clear in the surface topography image. The increased resolution of the smaller scan size is due to the number of data point per line being limited to 512 points per line and 512 lines per image. A UFM image of the same region at an excitation frequency of 759 kHz shown in Fig. 3b clearly shows the presence of the cracks. The grain size of the samples is in the range of a micron or less, but the crack measures several grains in length and runs along the grain boundaries.

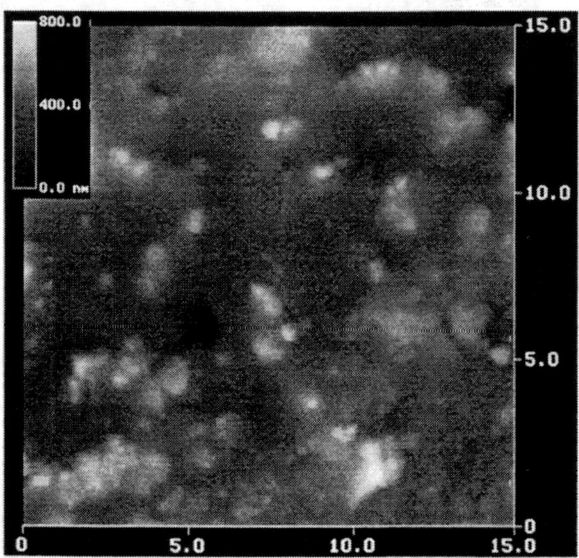


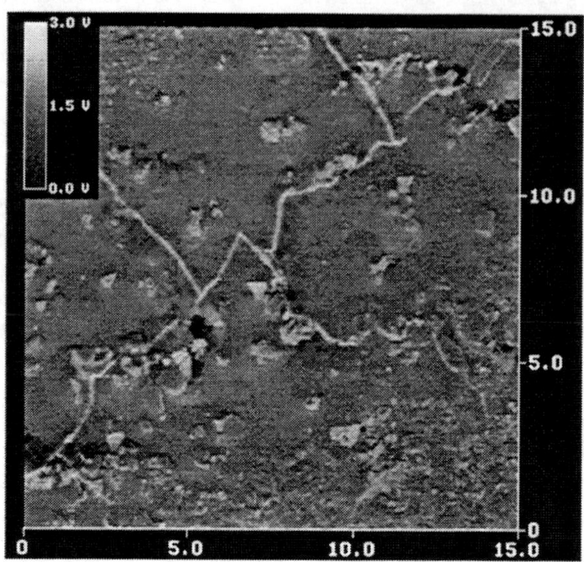


**Figure 3: a) On the top is an AFM topography image. b) On the bottom is a UFM amplitude image at an excitation frequency of 759 kHz. Scan area is 15 μm × 15 μm of crack line arrowed in Fig. 2.**

Figure 4 shows images of slider#2. The UFM images were acquired at 905 kHz. As it has been observed earlier the contrast of the UFM images is much better than AFM. The area marked in white on the AFM image does not exhibit enough contrast. On the other hand the UFM image 4b clearly shows crack like features or it could also be a grain loosened from the matrix. A magnified image of surface topography, 4a, and elasticity map, 4b, are shown in Fig. 5. While the contrast in AFM image is fair it does not reveal the features of cracking either or the grain structure of the material. The UFM image clearly shows long cracks running along the grain boundaries. The grain structure is quite clear.

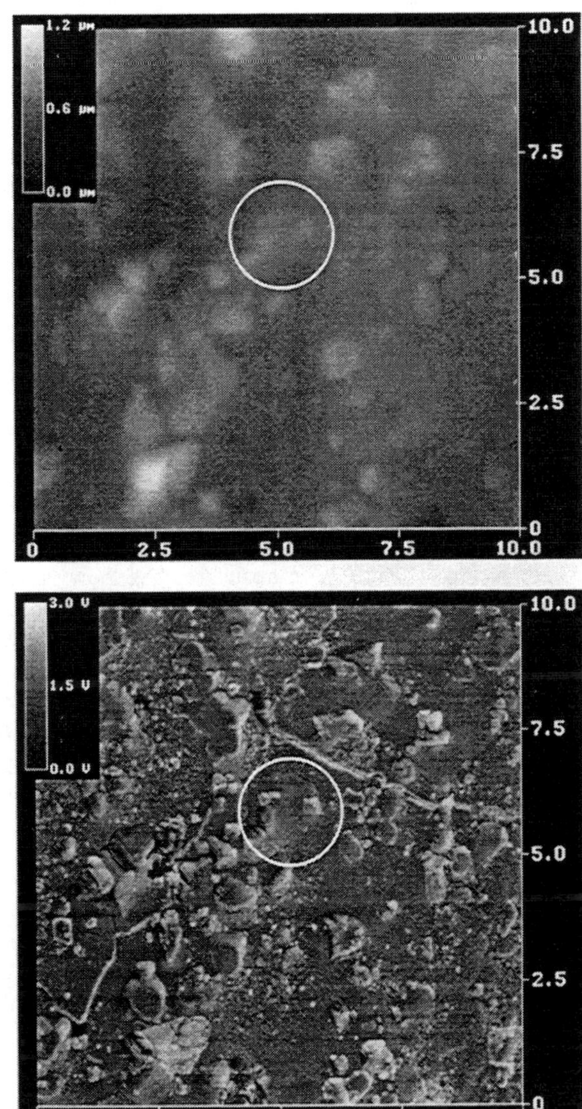


**Figure 4: a) On the top is an AFM topography image. b) On the bottom is a UFM amplitude image at an excitation frequency of 905 kHz. Scan area is 10 μm × 10 μm on slider #2.**

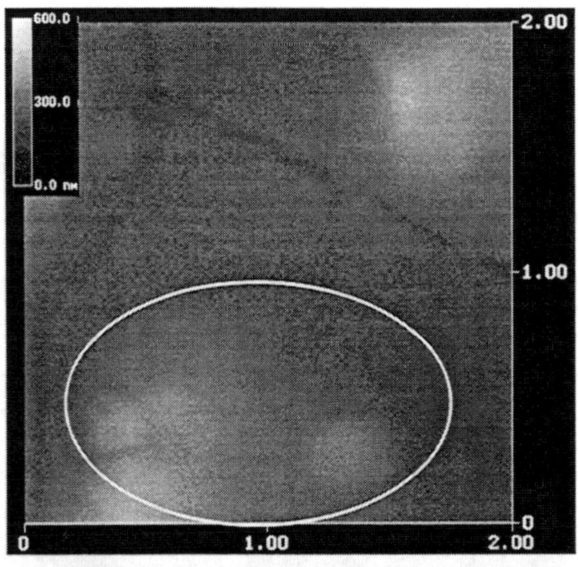


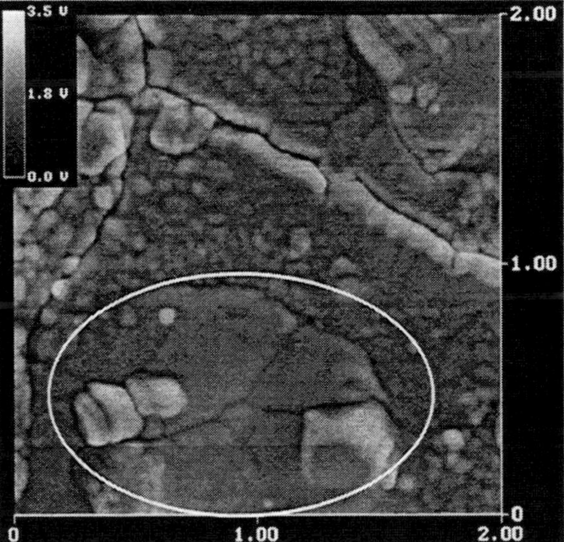


**Figure 5: A smaller scan of 2 μm on slider #2. a) On the top is an AFM topography image. b) On the bottom is a UFM amplitude image at an excitation frequency 905 kHz. The white circle encloses the structure show highlighted in previous Fig. 4.**

Figure 5, a magnified image of the region in white shows a region affected by the laser treatment. It looks as if a group of grains have loosened up from the matrix. An interesting feature in the cracks can also be observed. The area along side of the microcracks in the UFM appears to have a wavy texture to it. This seems to have been caused by thermal stresses of very fast melting and cooling due to the laser.

## UNTREATED REGION

For a comparison, AFM and UFM images were obtained in an untreated region of the sample. The images are shown in Fig. 6.

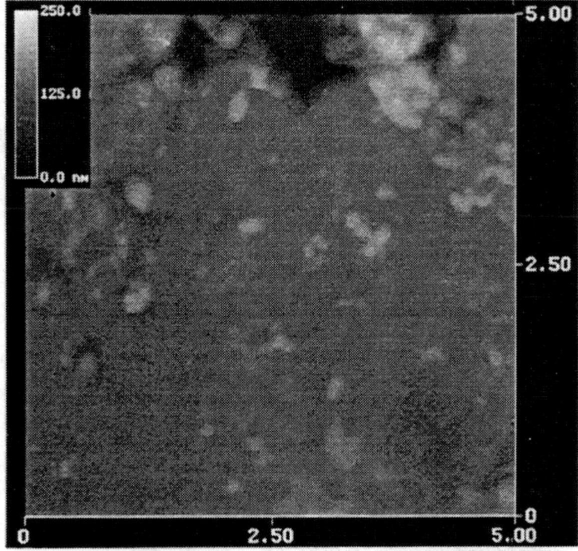


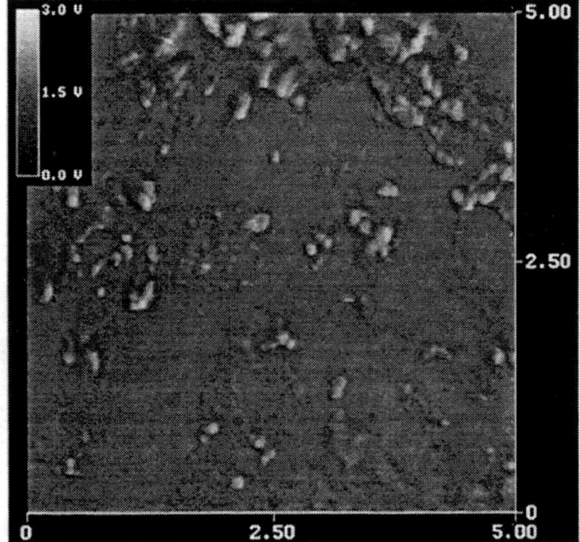


**Figure 6: a) on the top is an AFM topography image. b) On the bottom is a UFM amplitude image at an excitation frequency of 905 kHz. Scan area is 5 μm × 5 μm area on untreated Slider #3.**

Most features are common to both images and no indications of crack like features are observed. The microcracking was only found in areas that have been laser treated and was not found to have extended out of the laser treated area. Because the cracks did not extend out into the untreated area it was assumed that the cracks were formed by the thermal stress when the material resolidified after laser heating.

## DISCUSSION

The contrast in UFM images is due to variation in the elastic property of the material under the tip. The head slider material is a ceramic composite with the composition of $Al_2O_3$–TiC being about 70% aluminum oxide. Single crystal $Al_2O_3$ has hexagonal crystal structure and is elastically very anisotropic. In a polycrystalline material each grain could be of different orientation and hence can have a different elastic modulus. UFM images show large contrast variations and it is related to the variations in elastic modulus. In the UFM image, the elasticity recorded between the tip and the sample is the elasticity of near and subsurface region, which gives cracks a larger distinction.

Untreated regions of the head sliders do not exhibit any microcracking; while the laser treated regions show significant amount of cracks. The cracks run along the grain boundaries covering several grains. In many regions group of grains seemed to have gotten loosened from the matrix. Possible reasons for the damage and the microcracking, is the thermal stress caused by rapid increase in the temperature and rapid cooling of the laser treated regions. It is also interesting that most of the microcracks are observed running parallel to the laser line.

## CONCLUSIONS

Atomic Force Microscopy and Ultrasonic Force Microscopy have been performed on composite hard disc drive head sliders in laser treated and non-treated regions. AFM images show the topographic variations and anomalous regions of cracking are submerged in the height variation. UFM images clearly show the presence of microcracks in the midst of the grain structure in the laser treated regions. These images also reveal the microstructural inhomogeneities in the material. Micro-cracks running along the grain boundaries have been observed. Groups of grains loosened from the matrix due to laser treatment have also been observed. The microcracks appear to be running parallel to the laser line.

## ACKNOWLEDGMENTS


This research effort was supported by the National Science Foundation under Grant CMS-9905394 and by DARPA NDE-MURI under grant F49620-96-1-0442, the Center for Materials Diagnostic at the University of Dayton Research Center and also by the University of Dayton through the Department of Materials Engineering. The authors would like to thank Dr. Daniel Eylon and the personnel at the Center for Material Diagnostics along with Seagate Technology Corporation for their help and support.

## REFERENCES


1. Tam, A.C., "Laser processes for precise microfabrication of magnetic disk drive components", Riken Review, Vol. 32, pp. 71-76, 2001.
2. Chilamakuri, S. and Bhushan, B., "Contact analysis of laser textured disks in magnetic head-disk interface", Wear, Vol. 230, pp. 11-23, 1999.
3. Liu J.J., Li W., and Johnson K., "Current and Future approaches for laser texturing of thin film media", IEEE Transactions on Magnetics, Vol. 36, No. 1 pp. 125-131, 2000.
4. Lu Y.F., Song W.D., and Hong M.H., " Laser removal of particles from magnetic head sliders", Journal of Applied Physics, Vol. 80, No. 1, pp. 499-504, 1996.
5. Zhou, L., Kato, K., Umehara, N., and Miyake, Y., "Nanometre Scale island type texture with controllable height and area ratio formed by ion beam etching on hard disk head sliders", Nanotechnology, Vol. 10, pp. 363-372, 1999.
6. Zhou, L., Kato, K., Umehara, N., and Miyake, Y.," Friction and wear properties of hard coating materials on textured hard disk sliders", Wear, Vol. 243, pp. 133-139, 2000.
7. Wahl M. and Talke F., "Numerical simulation of the steady state flying characteristics of a 50% slider with surface texture", IEEE Transactions on Magnetics, Vol. 30, No. 6, pp. 4123-4125, 1995.
8. Zhang, J.H., Lee T.C., Ai X., and Lau, W.S." Investigation of the surface integrity of laser cut ceramics", Journal of materials processing technology, Vol. 57, pp. 304-310, 1996.
9. Bhushan, B., "Micro/nanotribology and its applications:" Kluwer Academic Publishers, Boston 1997.
10. Binning, G., Quate, C.F., and Gerber C.H. "Atomic Force Microscope", Physical review letters, Vol. 56, No. 9, pp. 930-933, 1986.
11. Maivald, P., Butt, H.J., Gould, SAC, Prater, C.B., Drake, B., Gurley, J.A., Elings, V.B. and Hansma, P.K..,"Using force modulation to image surface elasticities with the atomic force microscope", Nanotechnology, Vol. 2, pp. 103-106, 1991.
12. Poon, C. and Bhushan, B., "Nano-asperity contact analysis and surface optimization for magnetic head sliders/ disk contact", Wear, Vol. 202, pp. 83-98, 1996.
13. Gatzen, H., Ma, X., Scherge, M., Jhon, M.S., and Bauer, C.L. "Observations regarding the tribological properties of SiC and ALTiC Sliders", IEEE Transactions on Magnetics, Vol. 32, No. 5, pp. 3783-3785, 1996.
14. Koinkar, V. and Bhushan, B., "Microtribological studies of AL2O3, Al2O3-TiC, polycrystalline and single crystal Mn-Zn ferrite", Wear, Vol. 202, pp. 110-122, 1996.
15. Chekanov, A., Low, T., Alli, S., Liu, B., Teo, B.S., and Hu, S. "Microcracks of the Alumina of thin Film Head: Study and Simulation", IEEE Transactions on Magnetics, Vol. 31, No. 6, pp. 2991-2993, 1995.
16. Chekanov, A., Low, T., and Alli, S.," Microcracks of the Thin Film Head Alumina: L and U cracks", IEEE Transactions on Magnetics, Vol. 32, No. 5, pp. 3696-3698, 1996.
17. Yamanaka, K, "Ultrasonic Force Microscopy", MRS Bulletin, October 1996.
18. Burnham, N., Kulik, A.J., Gremaud, G., Gallo, P.J., and Oulevey, F. "Scanning local acceleration microscopy", Journal of Vacuum Science and Technology, Vol. 14, No. 2, pp. 794-799, 1995.
19. Burnham, N., Gremaud, G., Kulik, A.J., Gallo, P.J., and Oulevey, F. "Materials properties measurements: choosing the optimal scanning probe microscope configuration", Journal of vacuum science and technology, Vol. B-14, No. 2, pp. 1308-1312, 1996.
20. Rabe, U., and Arnold, W., "Acoustic Microscopy by atomic force microscopy", Applied Physics Letters, Vol. 64, No. 12, pp. 1493-1495, 1994.
21. Yamanaka, K. and Ogiso, H., "Ultrasonic Force Microscopy for nanometer resolution of subsurface imaging", Applied Physics Letters, Vol. 64, No. 2, pp. 178-180, 1994.

**Proceedings of**
**2001 ASME International Mechanical Engineering Congress and Exposition**
**November 11–16, 2001, New York, NY**
**NDE-Vol. 21**

**IMECE2001/NDE-25807**

# ULTRASONIC ASSESSMENT OF IMPACT DAMAGES IN FIBER-METAL LAMINATED PANELS

**Alexis M. Pierides, Yangxiong Liu, and Benjamin Liaw**
The City College of The City University of New York
Department of Mechanical Engineering
and
The CUNY Graduate School and University Center
Convent Avenue and 138th Street
New York, NY 10031

## ABSTRACT

Damages generated by instrumented drop-weight impacting onto various types of Glare fiber-metal laminated panels were assessed by the immersion ultrasonic NDE technique. Factors affecting impact damage tolerance of fiber-metal laminates, such as the choice of aluminum alloy, lay-up sequence, panel thickness, etc., were considered. One major goal of this study is to identify proper immersion ultrasound parameters for better detecting failure mechanisms, especially those hidden inside these opaque materials. They include B- or C-scanning, focused or flat transducers, transducer frequency, pulse-echo or through-transmission mode, etc.

## INTRODUCTION

Fiber-metal laminated panels are hybrid composites made of adhered alternating layers of thin metals (usually aluminum) and fiber-reinforced laminae. Commercially two types of fiber-metal laminates are available: Glare and ARALL, which employ glass-epoxy and aramid-epoxy, respectively. Since their development in late 1970's, they have been used in the aerospace and space industries [1-7]. Due to the hybrid nature of its two key constituents: metals and fiber-reinforced laminae, fiber-metal laminates provide better damage tolerance to fatigue crack growth, foreign object damage, and corrosion.

As reported in Ref. [8], low-energy-impact induced damage mechanisms in traditional composite panels have been studied quite extensively. Typically they are characterized by an incomplete penetration accompanied by various damage patterns, such as delamination, matrix cracking, fiber failure, etc. The exact failure mode and damage evolution sequence are greatly affected by several factors, such as constituent properties, lay-up configuration, thickness, bending rigidity, etc. On the other hand only a few studies have been reported in open literature on impact damages [3-5] and other thermomechanical behaviors [5-7] of fiber-metal laminates. Perhaps the closest studies are those reported on impact response of sandwich panels [9-11] The main objective of this study is to assess damages caused by low-energy impact in this type of relatively newer engineering composites using the immersion ultrasonic technique.

## SPECIMENS, IMPACT AND ULTRASOUND

Panels of 2024-T3 aluminum and various types of Glare were first cut into square specimens of 100 mm x 100 mm (or 4" x 4"). They were then clamped circumferentially along a diameter of 76 mm (or 3") in the specimen fixture and placed in an Instron-Dynatup 8520 pneumatic-assisted, instrumented drop-weight impact tester equipped with an environmental chamber (Fig. 1). A 16 mm- (or 5/8"-) diameter spherical impactor with a weight of 6.1 kg (or 13.4 lbs) was then dropped from a predetermined height to cause damage. Time histories of impact load were recorded by a PC-controlled high-speed A/D converter. After impact the specimen was then removed first for post-mortem inspection and photography. All edges of the damaged specimen were then sealed by silicon before being inspected with a PC-controlled immersion ultrasound system (Fig. 2).

## RESULTS AND DISCUSSION

A typical impact test result of this study is shown in Figs 3 and 4, which display damages and loading histories of panels made of (a) 0.063"-thick 2024-T3 aluminum alloy and (b) 0.056"-thick Glare 2, respectively, subject to a drop-weight impact of 30 J at room temperature. Glare 2 is a fiber-metal laminate formed by interlacing 2024-T3 aluminum sheets with unidirectional glass-epoxy laminae with a fiber volume fraction, $V_f$, of 50%. As shown in Figs 3 and 4, although the shape of loading histories was similar, different damage patterns were produced in these two panels. Severe damage, including partial punching through, was generated in the thicker and heavier aluminum panel; while only a minor indentation with a crack running along the fiber direction was induced in the thinner and lighter Glare 2 panel. The occurrence of impact damages is indicated by the abrupt drop in loading curves (Fig. 4), which also show that Glare 2, though thinner and lighter, can resist higher impact load before fracture.

Because of the deep indentation, ultrasonic assessment in the through-transmission mode (which employs a pair of flat transducers) produces more satisfactory results than those generated by the pulse-echo mode (which uses a single focused transducer). As shown in Fig. 5, the deep partially penetrated indentation occurred in the afore-mentioned 2024-T3 aluminum panel can be faithfully re-produced by C-scan ultrasound using a pair of 25 MHz through-transmission transducers.

Figure 6 shows the before and after impact C-scan ultrasonic through-transmission images generated by three pairs of flat transducers with different frequencies for the above mentioned Glare 2 panel. By comparing the before and after images, one can see that the dome-shaped impact site and the crack running along the fiber direction can be detected by the transducers with the frequency range (5 to 25 MHz) used in this study. The figure also indicates that transducers with higher frequency (e.g., 25 MHz) can generate ultrasonic images with better resolution.

The severity of impact damage (that is, the reduction in structural integrity) can be estimated easily with the composite B-scan images shown in Fig. 7. For instance, the acoustic impedance at the impact site was almost null after impact, indicating the catastrophic crushing effect on the future load-carrying capacity at that site. As shown in Fig. 7, the degree of this reduction in structural integrity was eased out gradually along the radial direction emanating from the impact site.

The second set of impact tests involved two types of Glare panels with different aluminum alloys. They are called Glare 1 and 2, which are fiber-metal laminates composed of alternating 7475-T76 and 2024-T3, respectively, aluminum alloy sheets of 0.012" with unidirectional glass-epoxy pre-preg layers ($V_f$ =50%, thickness: 0.01"). The total panel thickness is 0.056" for both Glare 1 and 2 panels. Figure 8 shows impact damage patterns in panels made of these two composites subject to different impact energies. As shown in Fig. 8(b), when impacted by a 20-J load, only plastic indentation but not fracture occurred in Glare 2 whereas cracks were generated in Glare 1 under the same impact load. It is clear that Glare 2, which incorporates the ductile but tougher 2024-T3, offers better impact resistance in comparison to Glare 1, which uses the stronger yet brittle 7475-T76.

The corresponding load-times histories of these tests are shown in Fig. 9. From this figure, one can conclude that when no fracture occurred, the load vs. time curves are fairly smooth other than some minor fluctuations (e.g., Glare 2, 20 J). This implies that excessive kinetic energy was mainly attenuated out through damping. On the other hand, for the cases with appearance of cracks, the load was drastically reduced; indicating the onset of fracture (e.g., Glare 1, 20 J). This phenomenon is similar to impact damage in aluminum/acrylic sandwich plates [9].

The third set of tests considered the effect of lay-up configuration in the prepreg. The results, as indicated by the impact damage patterns shown in Fig. 10, indicate that as far as resistance to impact load and crack formation is concerned, the quasi-isotropic configuration ([0°/45°/–45°/90°], Panel 14) has the best tolerance for impact damage with the cross-ply configurations ([0°/90°/90°/0°], Panel 5 and [45°/–45°/–45°/45°], Panel 13) being the second while the unidirectional configuration ($[0°]_4$, Panel 11) performing the worst. The corresponding impact loading histories are shown in Fig. 11.

Figure 12 shows the before and after impact C-scan images of the cross-ply [0°/90°/90°/0°] panel (3 aluminum + 2 glass-epoxy layers, total thickness: 0.076"). Although the panel's lay-up configuration is symmetric, the shape of impact damage seems to be stretched along the vertical 0° direction. On the other hand, when the same impact condition is applied to a thinner cross-ply [0°/90°] panel (3 aluminum + 2 glass-epoxy layers, total thickness: 0.056"), the shape of impact damage is more symmetric, as depicted in Fig. 13. This implies that thickness and detailed lay-up sequence affect the impact behavior in fiber-metal laminated panels.

The effect of number of layers was assessed for cross-ply [0°/90°/90°/0°] Glare panels with a total thickness varying from 0.044" to 0.172" were tested. The minimum impact energy required to cause cracking on the bottom face (opposite to the impact site) of the panel was measured. The result is shown in Fig. 14, which indicates that this minimum cracking energy is almost proportional parabolically to the thickness of the impacted panel. The corresponding impact loading curves are plotted in Fig. 15.

## CONCLUSIONS

From this study one can conclude that

- Impact damage tolerance in fiber-metal laminated panels is improved over that of aluminum; it can be further improved by using the ductile but tougher aluminum alloy (such as 2024-T3);
- Detailed information of impact damage in fiber-metal laminates can be revealed by using B- and C-scan ultrasound technique in through-transmission mode with higher transducer frequency;
- The shape of impact damage is affected by detailed lay-up sequence and total thickness of the impacted panel;
- Quasi-isotropic lay-up configuration provides the best impact resistance whereas unidirectional laminate offers the worst;
- The minimum impact energy required to cause cracking is almost proportional parabolically to thickness.

## ACKNOWLEDGEMENTS

This study was supported by NASA Faculty Award for Research (FAR) under Grant No. NAG3-2259 and by PSC-CUNY under Grants 61429-00 30 and 62466-00 31. Dr. Kenneth J. Bowles and Dr. John P. Gyekenyesi are the Technical Monitors of the NASA grant. Part of the equipment used in this investigation was acquired through Army Research Office Grant No. DAAD19-99-1-0366.

## REFERENCES

[1] Schijve, J., Van Lipzig, H. T. M., Van Gestel, G. F. J. A., and Hoeymakers, A. H. W., 1979, "Fatigue Properties of Adhesively-Bonded Laminated Sheet Materials of Aluminum Alloys," Engineering Fracture Mechanics, **12**, pp. 561-579.

[2] Marissen, R. and Vogelesang, L. B., 1981, "Development of a New Hybrid Material: ARALL," Proceedings, 2nd International SAMPE European Conference, Cannes, Frances, pp. 113-122.

[3] Vlot, A., 1993, "Impact Properties of Fiber Metal Laminates," Composite Engineering, **3**, pp. 911-927.

[4] Sun, C. T., Dicken, A., and Wu, H. F., 1993, "Characterization of Impact Damage in ARALL Laminates," Composites Science and Technology, **49**, pp. 139-144.

[5] Liaw, B. M., Liu, Y. X., and Villars, E. A., 2001, "Impact Damage Mechanisms in Fiber-Metal Laminates," Proceedings, SEM Annual Conference on Experimental and Applied Mechanics, Portland, OR, June 4-6, pp.536-539.

[6] Wu, H. F., 1989, "Statistical Analysis of Tensile Strength of ARALL Laminates," Journal of Composite Materials, **23**, pp. 1065-1080.

[7] Wu, H. F., 1990, "Temperature Dependence of the Tensile Properties of ARALL-4 Laminates," Journal of Materials Science, **25**, pp. 1120-1127.

[8] Abrate, S., 1998, *Impact on Composite Structures*, Cambridge University Press, Cambridge, UK.

[9] Liaw, B. M., Zeichner, G., and Liu, Y. X., 2000, "Impact Delamination and Fracture in Aluminum/Acrylic Sandwich Plates," Proceedings, SEM IX International Congress on Experimental Mechanics, Orlando, FL, June 5-8, pp. 515-518.

[10] Liaw, B. M., Pierides, A. M., and Liu, Y. X., 2001, "Ultrasonic Assessment of Impact Damage in Aluminum/Acrylic Sandwich Plates," Proceedings, SEM Annual Conference on Experimental and Applied Mechanics, Portland, OR, June 4-6, pp.264-268.

[11] Vaidya, U. K., Alimi, M., Mahinfalah, M., and Ulven, C., 2001, "Impact Response and NDE of Integrated Core Sandwich Composites," Proceedings, SEM Annual Congress on Experimental and Applied Mechanics, Portland, OR, June 4-6, pp. 352-353.

**Figure 1. Pneumatic-assisted instrumented drop-weight impact tester with environmental chamber.**

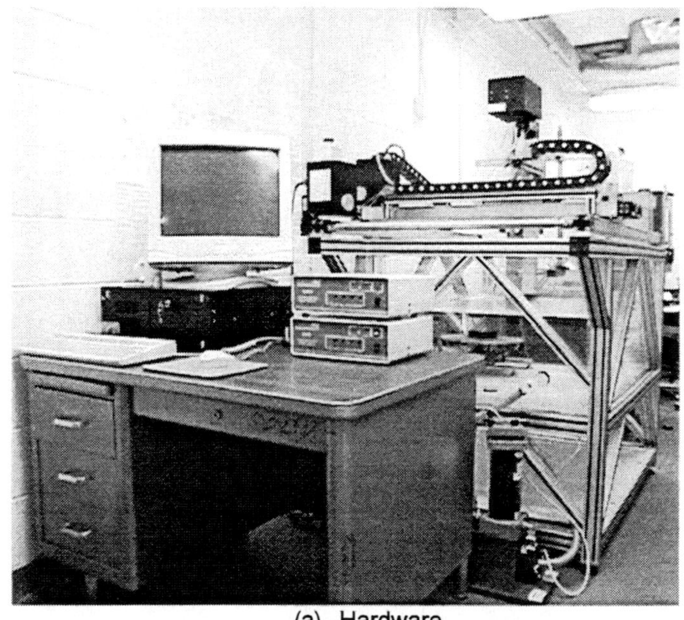

(a) Hardware

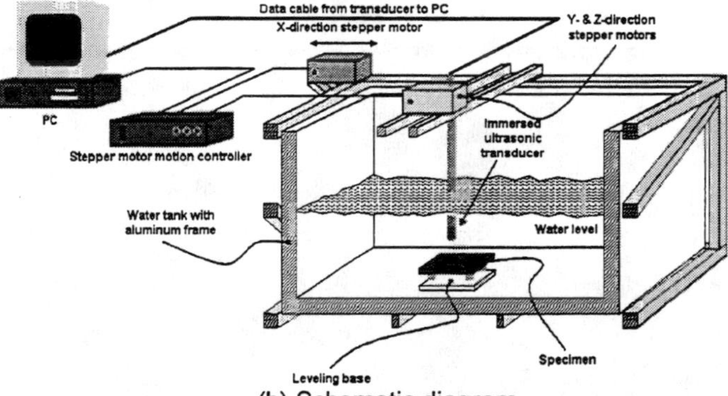


(b) Schematic diagram

**Figure 2. PC-controlled immersion ultrasound system.**

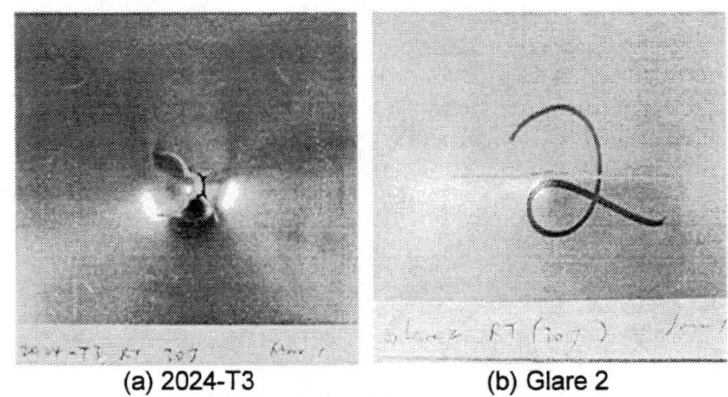

(a) 2024-T3 (b) Glare 2

**Figure 3. Damages in 2024-T3 aluminum and Glare 2 panels subject to a drop-weight impact of 30 J.**

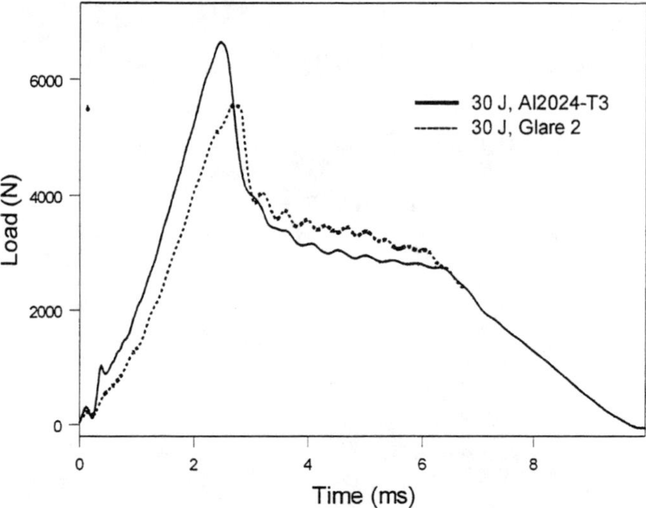


**Figure 4. Loading vs time histories of 2024-T3 aluminum and Glare 2 panels subject to a drop-weight impact of 30 J.**

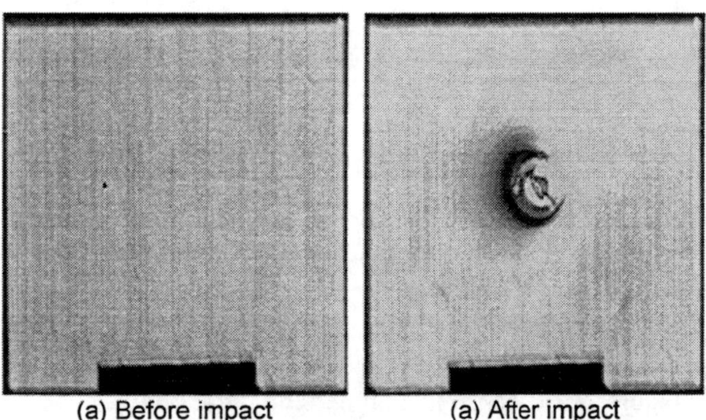

(a) Before impact (a) After impact

**Figure 5. Before and after impact ultrasonic C-scan images using 25 MHz through-transmission transducers of a 2024-T3 aluminum panel subject to a 30 J drop-weight impact.**

**Before Impact** **After Impact**

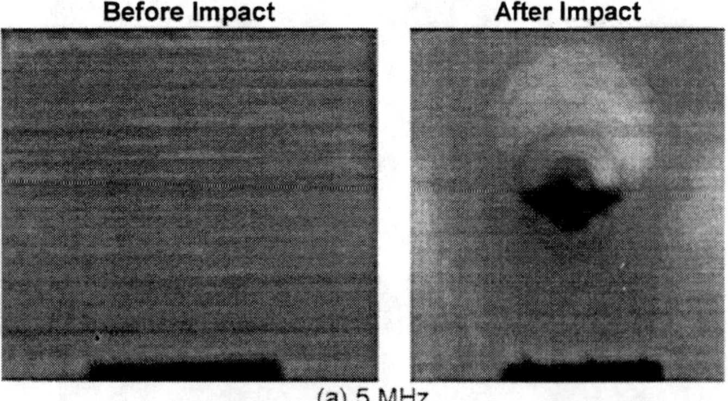

(a) 5 MHz

**Figure 6. C-scan ultrasonic through-transmission images of Glare 2 fiber-metal laminated panels produced by transducers of various frequencies.**

**Before Impact** **After Impact**

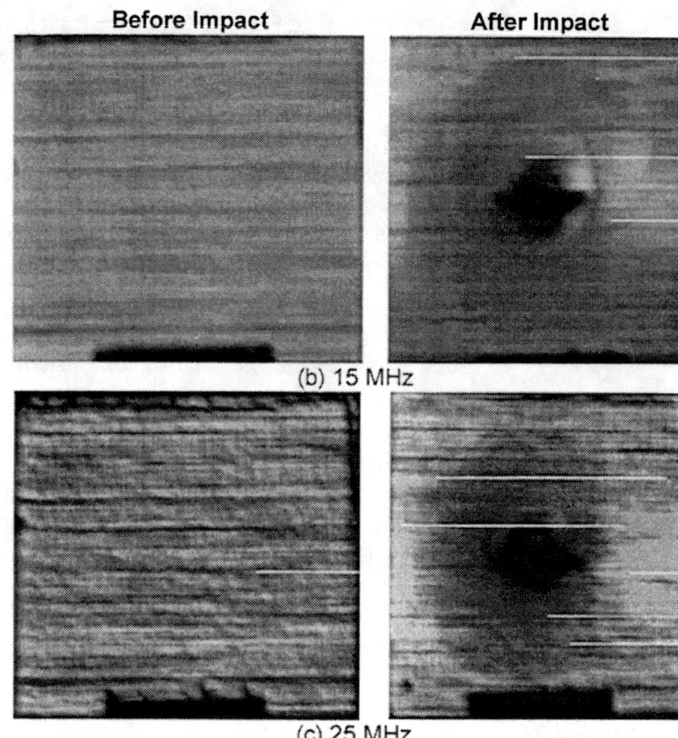

(b) 15 MHz

(c) 25 MHz

**Figure 6.** (Continued)

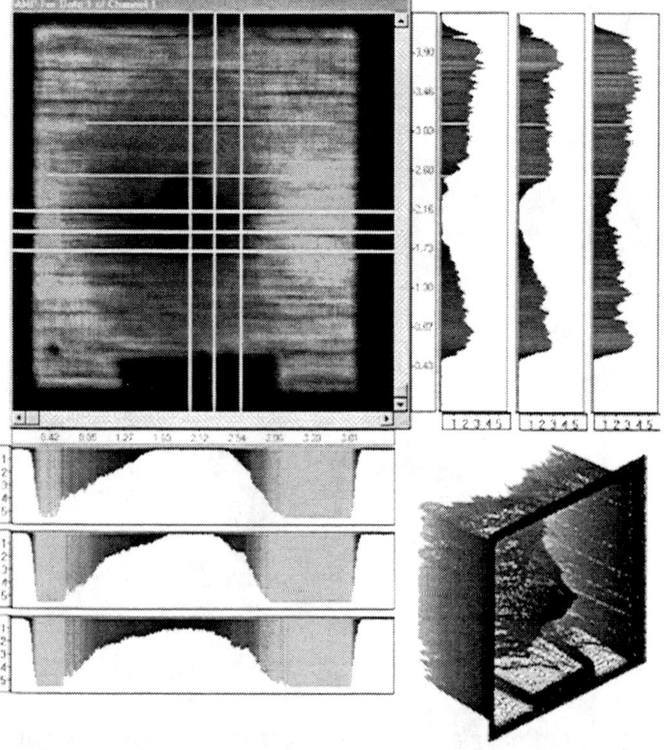

**Figure 7. B- & C- scan images of ultrasonic 25 MHz through-transmission of Glare 2 fiber-metal laminated panels subject to a 30 J drop-weight impact.**

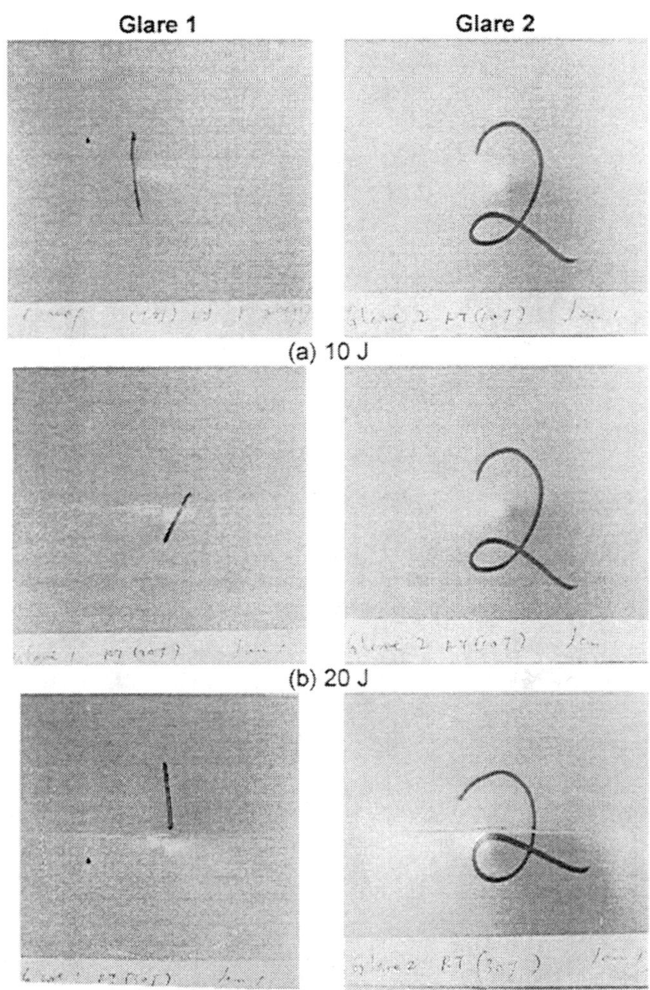


**Figure 8. Damage patterns in Glare 1 and 2 panels subject to drop-weight impacts of various impact energies.**

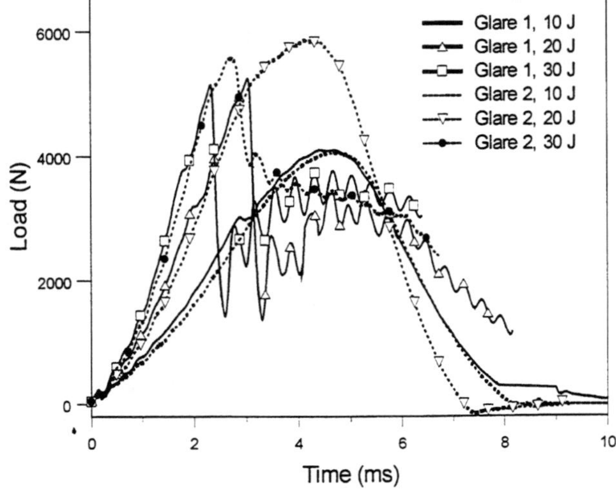


**Figure 9. Comparison of loading histories in Glare 1 and 2 panels under impact tests of various impact energies.**

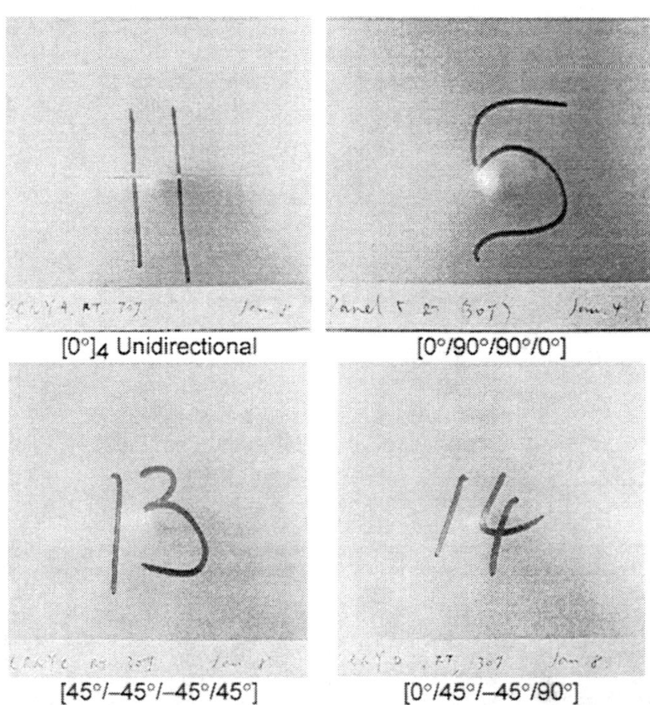


**Figure 10. Damage patterns in Glare panels of different lay-up configurations subject to a 30-J drop-weight impact.**

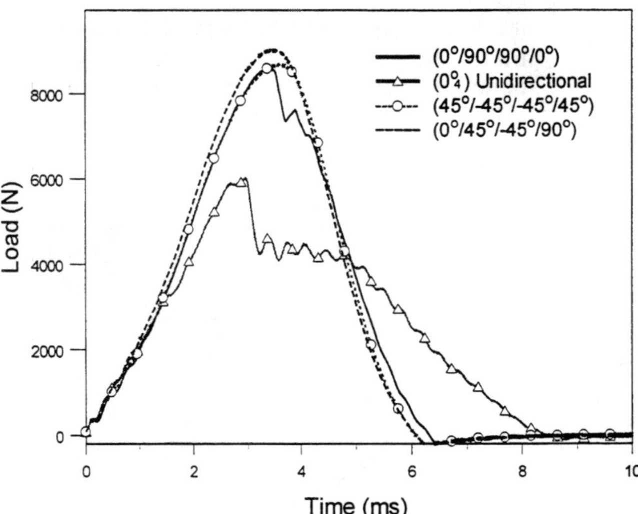


**Figure 11. Comparison of load-time histories Glare panels with different lay-up configurations subject to a 30-J drop-weight impact.**

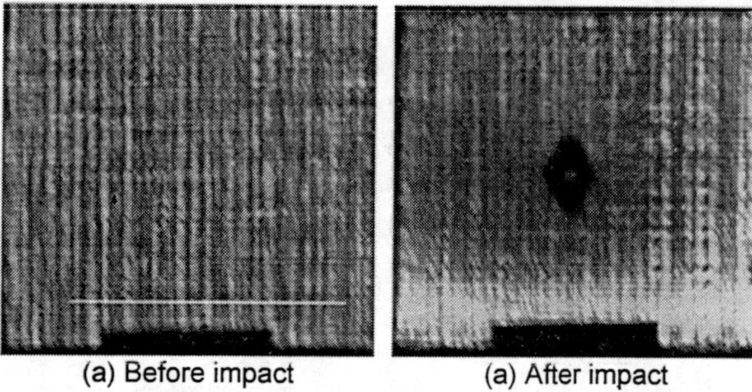


**Figure 12. Before and after impact ultrasonic C-scan images using 25 MHz through-transmission transducers of a Glare 5 panel ([0°/90°/90°/0°], 3/2 configuration, total thickness: 0.076") subject to a 30 J drop-weight impact.**

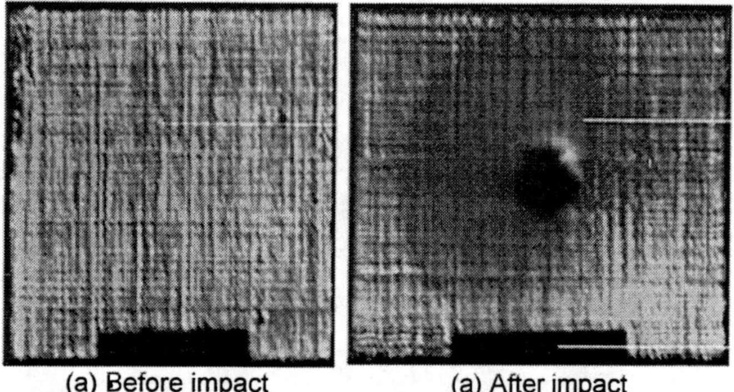


**Figure 13. Before and after impact ultrasonic C-scan images using 25 MHz through-transmission transducers of a Glare 3 panel ([0°/90°], 3/2 configuration, total thickness: 0.056") subject to a 30 J drop-weight impact.**

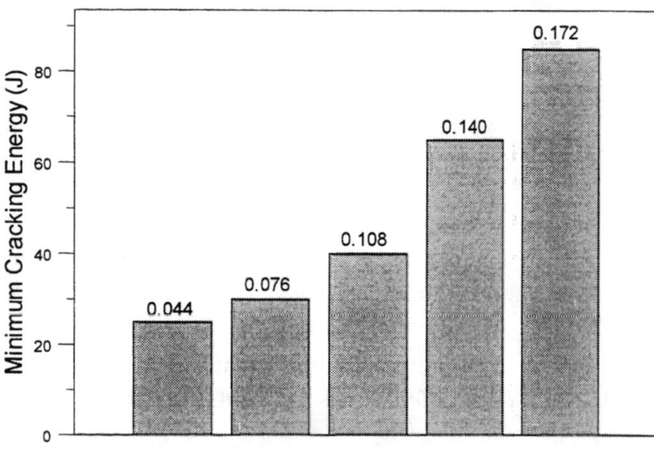


**Figure 14. Minimum impact energy required for cracking in cross-ply [0°/90°/90°/0°] Glare panels of different thickness.**

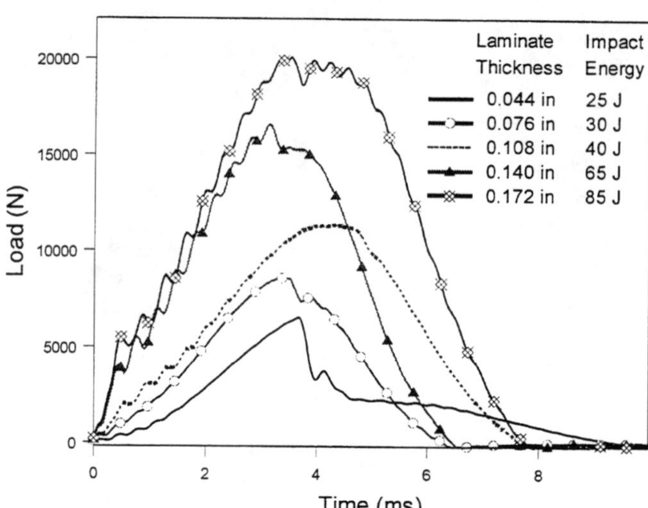


**Figure 15. Comparison of load-time histories for minimum impact energies required to cause cracking in cross-ply [0°/90°/90°/0°] Glare panels of different thickness.**

Proceedings of
2001 ASME International Mechanical Engineering Congress and Exposition
November 11–16, 2001, New York, NY
NDE-Vol. 21

**IMECE2001/NDE-25808**

# TRANSFER MATRIX FORMULATION WITH OPTICAL PENETRATION FOR AXISYMMETRIC THERMOELASTIC WAVE PROPAGATION IN FILMS

**Chen Li and Jiadao Lin**
Dept. of Mechanical and Aeronautical Engineering
Center for Advanced Materials Processing
Clarkson University, Potsdam, NY 13699-5725

**Cetin Cetinkaya**
Dept. of Mechanical and Aeronautical Engineering
Center for Advanced Materials Processing
Clarkson University, Potsdam, NY 13699-5725

## ABSTRACT

Using Laplace and Hankel integral transforms in time and the radial coordinate, a fully-coupled thermoelastic formulation based on the equation of motion and heat equation is developed to study the effects of axial optical penetration on axisymmetric wave propagation in thermoelastic layers and/or layered structures. It is demonstrated that the optical penetration has no effect on the entries of the sextic transfer matrix, however it introduces an equivalent forcing term for all state variables for both surfaces of a thermoelastic layer as opposed to the surface heating case in which the heating effect is localized in the heating volume (the thermal skin). The thickness of thermal skin depends on the light intensity modulation frequency while the optical penetration typically depends only on the wavelength of the light. This additional forcing vector is a function of the light intensity modulation frequency, the radial wave number, penetration decay rate, as well as thermoelastic material properties. Complexities in wavefields due to the nature of the forcing term are demonstrated and discussed. A thin copper layer with hypothetical penetration properties is considered for the demonstration of the current formulation.

## INTRODUCTION

The use of laser generated acoustic waves has found a wide range of applications from semiconductor metrology to structural health monitoring. This ultrasonic wave generation mechanism is a dynamic process of interacting elastic, thermal and optical effects. Laser ultrasonics can overcome some of the complications associated with contact ultrasonics such as the need for physical contact, the requirement for couplant gel or liquid, near perfect alignment requirements, or the flatness of surfaces. Furthermore, the deposition area of the photon energy can be small and localized, which could find various applications in nondestructive evaluation (NDE) of small-scale structures, and very high frequency excitations can be generated with aid of a pulsed laser. Using a pulsed or continuous wave (CW) laser, the photon energy can be deposited at a well-defined area as heat flux, and various patterns can also be generated for focusing and beam steering purposes. The level of deposited energy with a pulsed laser can be few order of magnitudes higher than that of a commonly used transducer and pulser-receiver unit. However, the conversion of light energy into heat flux and heat flux into elastic waves makes the process complicated, and considerably decreases the heat-mechanical energy conversion efficiency especially in the thermoelastic regime. The process of ablation has been utilized to increase the conversion efficiency, however it is normally a destructive process. Also, directivity of the elastic waves generated by a pulsed laser in the thermoelastic regime is different than that generated by a piezoelectric transducer. While the waves generated by a piezoelectric transducer are mostly pressure, the laser ultrasonics process generates mainly shear fields in the near field. The nature of the thermoelastic excitation is similar to that of applied shear field on a surface. Detection sensitivity and low energy conversion rate are often reported as two major issues in laser ultrasonics (Ref. [1]).

Since the early sixties, various applications of thermoelastic excitation have been reported (Ref. [1]). NDE of materials and structures has been a major application area. In addition, laser ultrasonics finds applications in a wide range of areas from semiconductor manufacturing to medical devices. For instance, a specific area is the generation of surface acceleration for the removal of fine particles (micron, nanometer level) from surfaces (Ref. [2]). Another area with great potential for laser ultrasonics is the testing and characterization of small-scale (micro) structures in which contact is very difficult to achieve. Using a laser with a nanosecond level pulse duration, micron-level spatial resolution could be achieved.

In this paper, for a thermoelastic layer, a fully coupled thermoelastic transfer matrix model with the axial optical penetration is introduced. Laplace and Hankel integral transforms in time and space are utilized as in (Ref. [3]). It is found that the optical penetration has no effect on the entries of the sextic transfer matrix. It, however, introduces an equivalent forcing term for all state variables (the displacement vector, stress components, heat and temperature) for both surfaces of a thermoelastic layer. In case of surface heating, the heating effect is localized in the heating volume (the thermal skin). The entries of the forcing vector are a function of the light intensity modulation frequency, the radial wave number, the axial penetration decay rate, as well as the thermoelastic material properties. Previous studies on the effect of the optical penetration by McDonalds (Ref. [4]) and Telschow and Conant (Ref. [5]) appear to aim at explaining the precursors observed in experimental data. The results, however, presented in Ref. [5] appears that the precursor is very close the noise floor of the measurement system, and that the bandwidth of the detector used in this study was *40 MHz*, while the pulsewidth of the laser used in the experiments was 10 nanoseconds. Furthermore, when one considers that the time dependency in the work of Telschow and Conant is assumed as a delta function, the spikes in the numerical solutions should be no surprise. In these two works, no reference to the $1/\omega^2$ displacement singularity in the integral transforms results is made. In a recent work, Enguehard and Bertrand (Ref. [6]) studied the one-dimensional wave problem by assuming neither thermal diffusion nor thermal propagation. The work by Enguehard and Bertrand (Ref. [6]) develops a deconvolution method for characterizing the laser pulse duration from experimental data for the one-dimensional case. With the current formulation, in addition to the effect of the penetration depth, the effect of the radial dependence of the beam intensity on the generation mechanism is also studied for axisymmetric thermoelastic propagation. It is found that the radial dependency of the laser heat distribution can play an important role in generating longitudinal waves in a layer or a layered structure.

## TRANSFER MATRIX FORMULATION

A typical way of representing the optical penetration is based on the observation that in solids the intensity of light decays exponentially in the direction of propagation (Ref. [1]) as the optical power is absorbed by the medium. In cylindrical coordinates, this behavior results in a volumetric thermal excitation field specified by

$$h(r,z,t) = A\, e^{-\alpha z}\, r_o(t, r)$$

per unit volume where the coordinate $z$ is in the axial (propagation) direction, $A$ the amplitude of the light intensity, $\alpha$ the absorption coefficient, $r_o(t, r)$ represents the temporal ($t$) and radial ($r$) dependency. In generating a first principle-based transfer matrix formulation, the linearized governing equations for the displacement, $\vec{u}$, and absolute temperature, $T$, fields for a homogeneous, isotropic medium consist of two coupled partial differential equations; the (Navier's) equation of motion and energy equation are utilized (Ref. [8]):

$$(\lambda+2\mu)\nabla(\nabla.\vec{u}) - \mu\nabla\times(\nabla\times\vec{u}) - \beta\nabla T + \rho_o(\vec{f}-\ddot{\vec{u}}) = 0 \qquad (1)$$

$$-\kappa\nabla^2 T + T_o\beta\nabla.(\dot{\vec{u}} + \tau_o\ddot{\vec{u}}) + \rho_o\gamma(\dot{T} + \tau_o\ddot{T}) - h(r,z,t) = 0 \qquad (2)$$

where $\lambda$ and $\mu$ are Lamé constants of the layer material, $\rho_o$ the layer material density, $\kappa$ the thermal conductivity, $\gamma$ the specific heat capacity, $h(r,z,t)$ the internal heat source intensity per unit volume, $T_o$ the temperature at the normal state, $\tau_o$ the relaxation time, and overdot represents time-derivative, and $\beta = (3\lambda+2\mu)\alpha$ the thermo-elastic coupling coefficient where $\alpha$ is the thermal expansion coefficient. The relaxation time $\tau_o$ "represents the time-lag needed to establish steady-state heat conduction in an element of volume when a temperature gradient is suddenly imposed on that element" (Ref. [8]). These coupled governing equations along with the following constitutive relation and Fourier's law for isotropic materials form the basis for the current fully-coupled formulation:

$$\sigma_{ij} = -\beta T\delta_{ij} + \lambda u_{n,n}\delta_{ij} + \mu(u_{i,j} + u_{j,i}) \qquad (3)$$

$$q_i + \tau_o\dot{q}_i = -\kappa T_{,j} \qquad (4)$$

where $\sigma_{ij}$ is the *ij-th* component of the stress tensor, $u_i$ the *i-th* component of the displacement vector, $q_i$ heat flux in the direction *i*, $\delta_{ij}$ the Kronecker delta, and the index following the comma in the subscript indicates differentiation with respect to corresponding coordinate. This model, also referred to as the generalized dynamic theory of linear thermoelasticity, was first proposed by Lord and Shulman (Ref. [8]). The classical heat equation is a parabolic partial differential equation, for which only diffusion process can be modeled. In the generalized theory the energy equation Eq. (2) contains a term with $\ddot{T}$ which makes it a hyperbolic equation as the displacement equation Eq. (1). Changing the nature of equations solves the problem of infinite speed thermal wave propagation predicted by the heat diffusion equation. The propagation of thermal waves, the second sound effect, was first postulated by Maxwell in 1867 on the physical argument that heat pulses can not propagate with infinite velocity in matter.

In Ref. [9] by Cetinkaya, et al., based on the generalized dynamic theory of thermoelasticity, a transfer matrix formulation including the second sound effect is developed for axisymmetric propagation in a thermoelastic layer by utilizing a double integral transform method. The second sound effect, through a relaxation time term, is included. Similar transfer function formulations are utilized in Ref. [3] by Sve and Miklowitz for calculating stress components due to laser beam excitation without considering a relaxation term, and in Ref. [10] for surface displacement calculations under laser heating. The transfer matrix formulation for axisymmetric elastic waves (without optical penetration) was given in Ref. [11]. The transfer matrix formulation has been utilized for wave propagation problems in various areas since its introduction (Refs. [12] and [13]) in seismology. The reflectivity and transmission as well as the localization of thermoelastic waves in layered structures are demonstrated in Ref. [14].

In generating a transfer matrix formulation for a longitudinal wave propagation study, the Laplace transform (with respect to the scaled time $\tau = c_T\, t/H$) of equations (1) and (2) along with equations (3) and (4) in the displacement potential function and temperature are performed in one-dimension. Both components ($\varphi$ and $\psi$) of the vector potential are needed in axi-symmetric analysis. The zero-th and first order Hankel transforms of the potential functions ($\varphi$ and $\psi$), respectively, for the scaled axial coordinate $\xi = z/H$ are applied to the wave equations in potentials. The details of the derivations can be found in Ref. [2]. Assuming that the thermoelastic state is initially ($\tau = 0$) at rest, a set of inhomogenous differential equations are the result of these potential functions and the temperature. The solutions to these inhomogenous differential equations with the internal distributed heat source $h(r, z, t)$ are obtained as

$$\overline{T}^o = A_1 C_1(p,k) e^{-c_1\xi} + A_2 C_1(p,k) e^{+c_1\xi} + A_3 C_2(p,k) e^{-d_1\xi} + A_4 C_2(p,k) e^{+d_1\xi} - \frac{e^{-\alpha\xi}(\alpha^2 + b_1)\rho(p,k)}{b_2(D_1 + D_2\alpha^2 + D_3\alpha^4)} \quad (5)$$

$$\overline{\Phi}^o = A_1 e^{-c_1\xi} + A_2 e^{+c_1\xi} + A_3 e^{-d_1\xi} + A_4 e^{+d_1\xi} + \frac{e^{-\alpha\xi}\rho(p,k)}{D_1 + D_2\alpha^2 + D_3\alpha^4} \quad (6)$$

$$\overline{\Psi}^1 = B_1 e^{-h\xi} - B_2 e^{+h\xi} \quad (7)$$

where $B_1$, $B_2$, $A_1$, $A_2$, $A_3$, and $A_4$ are the integration constants,

$$c_1 = \frac{1}{\sqrt{2a_3}}\left(-a_1 - a_3 b_1 + a_2 b_2 - \sqrt{-4a_3 a_1 b_1 + (a_1 + a_3 b_1 - a_2 b_2)^2}\right)^{1/2}$$

$$d_1 = \frac{1}{\sqrt{2a_3}}\left(-a_1 - a_3 b_1 + a_2 b_2 + \sqrt{-4a_3 a_1 b_1 + (a_1 + a_3 b_1 - a_2 b_2)^2}\right)^{1/2}$$

$$h^2 = p^2 + k^2$$

$$D_1 = a_2 + \frac{a_1 b_1}{b_2}$$

$$D_2 = a_3 + \frac{a_1}{b_2} + \frac{a_4 b_1}{b_2}$$

$$D_3 = \frac{a_4}{b_2}$$

$$a_1 = \frac{\kappa k^2}{H^2} + \rho_0 \gamma \frac{c_T}{H} p$$

$$a_2 = -T_0 \beta \frac{c_T}{H^2} p k^2$$

$$a_3 = T_0 \beta \frac{c_T}{H^2} p$$

$$a_4 = -\frac{\kappa}{H^2}$$

$$b_1 = -\frac{p^2 + k^2}{a^2}$$

$$b_2 = -\frac{H\beta}{c_L^2 \rho_0}$$

$$C_1(p,k) = -\frac{(b_1 + c_1^2)}{b_2}$$

$$C_2(p,k) = -\frac{(b_1 + d_1^2)}{b_2}$$

$\Phi = \varphi/H$, $\Psi = \psi/H$ the scaled scalar potential functions, $H$ is a characteristic distance (in the following sections, $H$ will be taken as the thickness of a layer), respectively, $p$ and k are the Laplace and Hankel transform variables, the overbar represents a Laplace transformed field variable, the superscript 0 and 1 indicate the order of the Hankel transform of the variable, and $a = c_L/c_T$, $c_L$ and $c_T$ are the propagation speeds of irrotational isothermal and equivoluminal isothermal waves in an isotropic medium, respectively. From the transformed potentials and the temperature, the displacement and normal stress component as well as heat flux can be calculated in the transformed domain. Substituting the scalar potential and temperature fields (Eqs. (5), (6) and (7)) into the transformed constitutive relation and Fourier's Law from (Eqs. 3 and 4), a matrix representation for the integration coefficients $A_1$, $A_2$, $A_3$, and $A_4$ results in the form of

$$[T(\xi)]\{B_1 \;\; B_2 \;\; A_1 \;\; A_2 \;\; A_3 \;\; A_4\}^T = \{d(\xi)\}^T + \{\overline{u}^1_\rho \;\; \overline{u}^o_{\xi\xi} \;\; \overline{\sigma}^o_{\xi\xi} \;\; \overline{\sigma}^1_{\xi\rho} \;\; \overline{T}^o \;\; \overline{q}^o_\xi\}^T \quad (8)$$

where the sextuple vector $\{d(\xi)\}$ is obtained as

$$d(\xi)=\begin{bmatrix} -\dfrac{e^{-\alpha\xi}k\rho(p,k)}{D_1+\alpha^2 D_2+\alpha^4 D_3} \\ -\dfrac{e^{-\alpha\xi}\alpha\rho(p,k)}{D_1+\alpha^2 D_2+\alpha^4 D_3} \\ \dfrac{e^{-\alpha\xi}\left(H\alpha^2\beta+H\beta b_1+\left(-k^2\lambda+\alpha^2(\lambda+2\mu)\right)b_2\right)\rho(p,k)}{H\left(D_1+\alpha^2 D_2+\alpha^4 D_3\right)b_2} \\ \dfrac{2e^{-\alpha\xi}k\alpha\mu\rho(p,k)}{H\left(D_1+\alpha^2 D_2+\alpha^4 D_3\right)} \\ -\dfrac{e^{-\alpha\xi}\left(\alpha^2+b_1\right)\rho(p,k)}{\left(D_1+\alpha^2 D_2+\alpha^4 D_3\right)b_2} \\ -\dfrac{e^{-\alpha\xi}\alpha\kappa\left(\alpha^2+b_1\right)\rho(p,k)}{H^2\left(D_1+\alpha^2 D_2+\alpha^4 A_3\right)b_2\left(1+p\tau_0\right)} \end{bmatrix} \quad (9)$$

where the superscript $T$ stands for the transpose operation. A transfer matrix for a single layer defined in the spatial interval $\xi$ in [0, 1] is generated by eliminating the integration coefficients by specifying the displacement, stress, temperature and heat flux on both surfaces of the layer:

$$[T(0)]\{B_1 \quad B_2 \quad A_1 \quad A_2 \quad A_3 \quad A_4\}^T = \{d(0)\}^T + \{\overline{u}_\rho^1 \quad \overline{u}_{\xi\xi}^o \quad \overline{\sigma}_{\xi\xi}^o \quad \overline{\sigma}_{\xi\rho}^1 \quad \overline{T}^o \quad \overline{q}_\xi^o\}^T \quad (10)$$

$\xi = 0$ for the top of the layer, and

$$[T(1)]\{B_1 \quad B_2 \quad A_1 \quad A_2 \quad A_3 \quad A_4\}^T = \{d(1)\}^T + \{\overline{u}_\rho^1 \quad \overline{u}_{\xi\xi}^o \quad \overline{\sigma}_{\xi\xi}^o \quad \overline{\sigma}_{\xi\rho}^1 \quad \overline{T}^o \quad \overline{q}_\xi^o\}^T \quad (11)$$

$\xi = 1$ for the bottom of the layer. The elimination of the integration constants in these two boundary conditions yields the following matrix equation:

$$[T(1)][T(0)]^{-1}\left(\{\overline{u}_\rho^1 \quad \overline{u}_{\xi\xi}^o \quad \overline{\sigma}_{\xi\xi}^o \quad \overline{\sigma}_{\xi\rho}^1 \quad \overline{T}^o \quad \overline{q}_\xi^o\}_R^T + \{d(1)\}^T\right) = \left(\{\overline{u}_\rho^1 \quad \overline{u}_{\xi\xi}^o \quad \overline{\sigma}_{\xi\xi}^o \quad \overline{\sigma}_{\xi\rho}^1 \quad \overline{T}^o \quad \overline{q}_\xi^o\}_L^T + \{d(0)\}^T\right) \quad (12)$$

This form indicates that equivalent thermoelastic states at the two surfaces can be determined for any type of the optical penetration. The penetration effect is mapped to thermoelastic states at the boundaries. The deposited energy (for pulsed lasers) and/or power (for CW lasers) in the layer come into the dynamics through the equivalent states at the boundaries. The effect of the optical penetration on the excitation of a thermoelastic layer in Eq. (12) is apparent. As it will become clear from the following discussions since all state variables are excited by $\{d(\xi)\}$, the resulting wave fields could become complicated. For example, high waveguide modes can be excited even at low excitation frequencies. So it is reasonable to expect that the resulting near-field wave field will be elaborate as higher order plate modes are excited due to the optical penetration.

## EQUILIVENT EXCITATION STATES

The optical penetration into a pure copper thin film is considered. The thermoelastic properties of copper are listed in Table 1.

| | |
|---|---|
| $\lambda$ = 130.9 Gpa | *Lamè Coefficient* |
| $\mu$ = 46.2 Gpa | *Lamè Coefficient* |
| $\rho$=8833.0 kg/m$^3$ | *Material Density* |
| $c_L$= 5027.9 m/s | *Speed of Irrotational Isothermal Wave* |
| $c_T$= 2287.0 m/s | *Speed of Equivoluminal Isothermal Wave* |
| $H$=12.7 10$^{-6}$ m | *Layer Thickness* |
| $\alpha$= 16.0 10$^{-6}$ 1/°C | *Thermal Expansion Coefficient* |
| $\kappa$=401.0 W/(m °C) | *Conductivity* |
| $\gamma$=385.0J/(kg °C) | *Specific Heat* |
| $T_0$ = 300.2 K | *Initial Temperature* |
| $\tau_0$ = 1.0 10$^{-7}$ s | *Relaxation Time Coefficient* |

***Table 1.*** Thermal and mechanical properties of pure copper at the specified initial temperature $T$=$T_0$.

The opto-thermal loading is in the form of

$$h(r,z,t) = f(t)\, g(r)\, h_o(z).$$

In case of the exponential optical absorption,

$$h_o(z) = exp(-\alpha z)$$

while $g(r) = H(r\text{-}H)$ where $H(.)$ is Heaviside function. In the current study, a harmonic thermal loading in time is applied by modulating the light intensity at radial frequency $\omega_o$;

$$f(t) = A\, Sin(\omega_o t).$$

For a more realistic laser loading the average power deposition should also be taken into account, which results in

$$f(t) = A(1+Sin(w_o t)).$$

Since the focus of this study is on the dynamics effects rather than much slower diffusion process, the component $\omega_o$=0 corresponding to the static thermal loading is neglected for simplicity. Penetration depth is typically a function of wavelength of light, and can be controlled by its tuning. The penetration depths considered in the following calculations for copper are hypothetical and are for demonstration purposes only. The deep light penetration could be achieved in relatively thick layers of polymers and ceramics easier than metals. The objective here is to emphasize the effect, rather than to study the optoacoustic response of copper. A single layer of copper with the thickness of *12.7* micrometers is considered.

The effects of penetration depth and the excitation of the periodic intensity light source are considered. The concepts of the optical penetration exponent $n_\alpha$ and the radial wavelength exponent $n_\beta$ are introduced. The absorption coefficient and radial wavelength in scaled coordinates (normalized in the layer thickness *H*) are given as $\alpha = 10^{n_\alpha}$ and $\lambda = 10^{n_\beta}$, respectively. The two excitation frequencies with respect to the longitudinal resonance frequency of the thin film are examined, namely, a low frequency ($0.1\ f_r$), and the layer resonance frequency ($f_r = 197.95\ MHz$). To clarify the effects of the optical penetration, the amplitudes of the components of the equivalent excitation vector (Eq. 9) are plotted for $n_\alpha$ and $n_\beta$ in the intervals [-3, 3] and [3, -2], respectively. As $n_\alpha$ tends to a large positive number, the optical penetration becomes shallower. As $n_\alpha$ tends to a small negative number, the penetration becomes deeper and the layer eventually becomes transparent. Large $n_\beta$ corresponds to a long radial wave. The larger the value of $n_\beta$ is, the closer the wave propagation problem becomes the one-dimensional case. The deep and shallow penetration regions in the $n_\alpha$ line and the long/short radial waves regions in the $n_\beta$ line are depicted in Fig. 1.a. In the current study, only the left-hand side of the film from which direction the laser light penetrates into the layer material is considered. In case of deep penetration, the right-hand side of the film should also be considered. The axisymmetric problem becomes a one-dimensional problem as the wavelength in the radial coordinate becomes large. Figs. 1 through 4 reflect this behavior. For $n_\beta >$ 1, the 'characteristic of the propagation becomes one-dimensional.

In Figs. 1-4, the equivalent power flux of a modulated 5.0 Watt CW laser with a circular beam and beam diameter *H* (the layer thickness = 12.7 μm) is considered. The radial displacement component $u_r$ at the low modulation frequency is mostly affected in the long radial wavelength/ the deep penetration domain defined by the intervals $n_\beta \in [0.5, 2]$ and $n_\alpha \in [0, -\infty]$ at the low modulation frequency excitation (Fig. 1.a). The base of the amplitude surface becomes narrower as the optical penetration becomes shallower and/or the radial wavelength gets shorter. Around the point $n_\beta \approx -1/2$ and $n_\alpha \approx 1$, a singularity appears alone a linear line towards the point $n_\beta \approx -2$ and $n_\alpha \approx 3$ of the plot. The singularity in the short wave/the shallow penetration depth limit seems unaffected by the modulation frequency. The line seems to extend to $n_\alpha \approx \infty$ (the long wave limit) and $n_\beta \approx -\infty$ (the deep penetration limit). At the resonance modulation frequency, the deep penetration/the long radial wavelength domain for $u_r$ shrinks compared to the lower frequency response and the maximum of the amplitude surface are shifted towards the line $n_\beta \approx 1/3$ (Fig. 1.b). The singularity behavior essentially remains unchanged. This is because they share the same denominator as presented in Eq. (9). Both displacement components are in the order of nanometers. It is also observed that as the frequency increases, the magnitude of the displacement $u_r$ tends to decrease approximately by a few orders of magnitude for each frequency increment (Fig. 1.a and b).

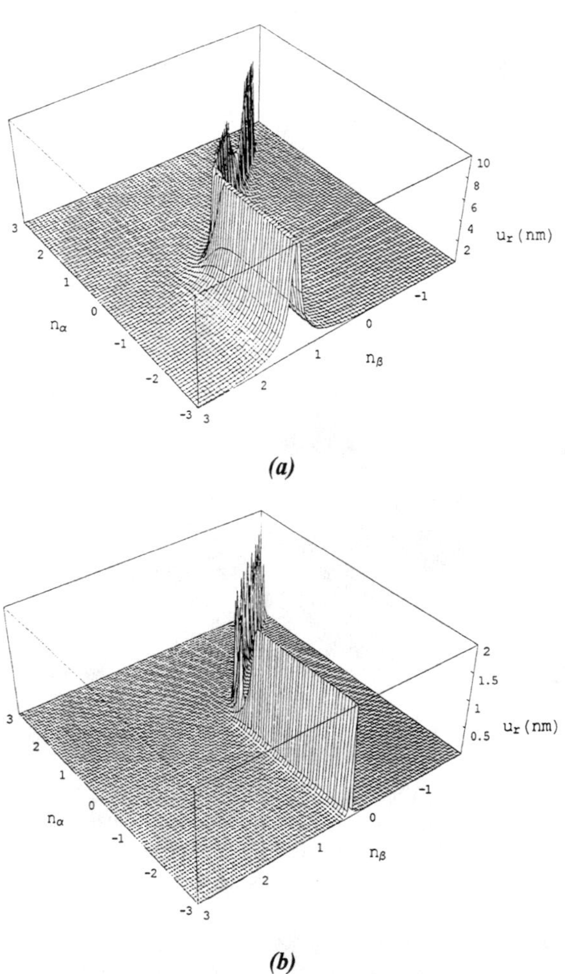


***Figure 1.*** The amplitudes of the radial displacement component $u_r$ in the unscaled coordinates as a function of the optical penetration exponent $n_\alpha$ and the radial wavelength exponent $n_\beta$ at ***(a)*** a low frequency ($0.1\ \omega_r$) and ***(b)*** the resonance frequency ($\omega_r$)

The magnitude of the axial displacement component $u_z$ in the $n_\alpha$- $n_\beta$ plane is similar to that of $u_r$ except for its long wave behavior. Since the propagation problem becomes one-dimensional at the long wave limit, while the component $u_r$ disappears, $u_z$ becomes dominant. In the shallow penetration range, the $u_z$ contribution is minimal at the low frequency.

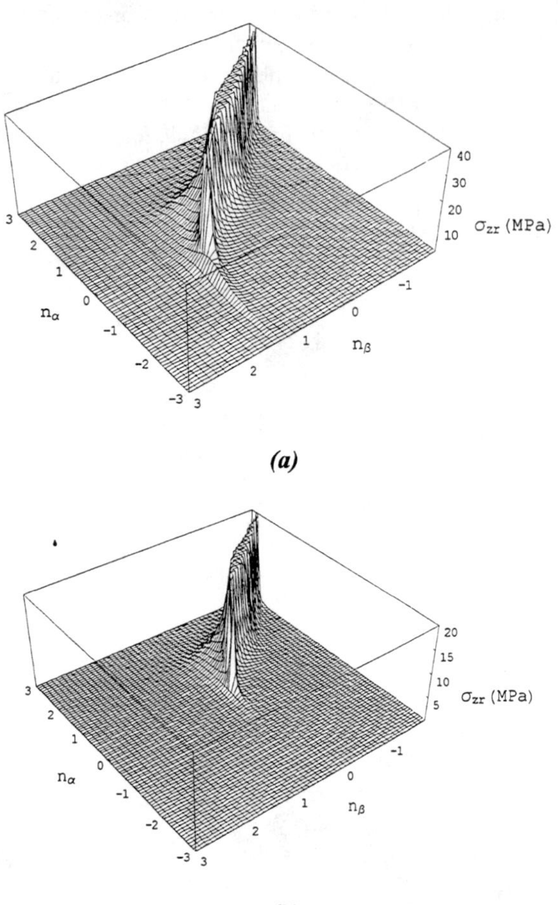


***(b)***

***Figure 2.*** The amplitudes of the stress component $\sigma_{zr}$ in the unscaled coordinates as a function of the optical penetration exponent $n_\alpha$ and the radial wavelength exponent $n_\beta$ at ***(a)*** a low frequency ($0.1\ \omega_r$) and ***(b)*** the resonance frequency ($\omega_r$).

The amplitude surfaces for the stress components $\sigma_{zr}$ are presented in Fig. 2.a and b at the three modulation frequency levels. The overall forms of the surfaces for $\sigma_{zz}$ and $u_z$ are of similarities in their limiting behaviors. In the same manner, the surfaces for $\sigma_{zz}$ and $u_z$ are similar. In the deep penetration/the long wave domain, $\sigma_{zr}$ is much less affected by the change in the modulation frequency that $\sigma_{zz}$. The stress amplitudes are on the order of MPa. In general, $\sigma_{zz}$ is larger than $\sigma_{zr}$ in the deep penetration/the long wave domain. In the short wavelength/the shallow penetration zone, it appears that the limiting behaviors of the two stress components are rather similar. As the propagation becomes one-dimensional (for large values for $n_\beta$), the shear components fades and the axial normal stress reaches to a limit value. This observation leads to the conclusion that the utilization of optical penetration and the distribution of the radial dependency of the laser beam can result in relatively strong longitudinal wavefronts in certain regions of the $n_\alpha$ and $n_\beta$ plane. As in the displacements, the amplitudes of the stress components significantly decrease.

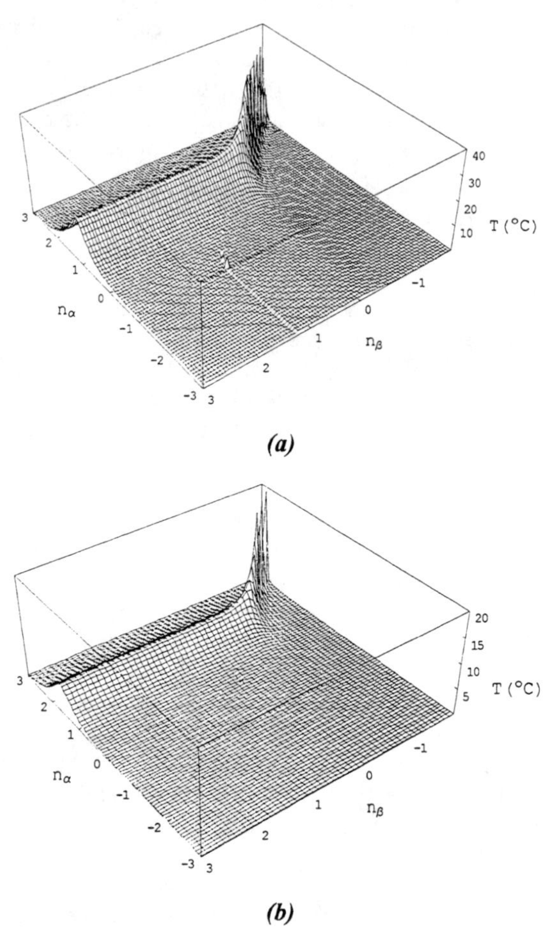


***(b)***

***Figure 3.*** The amplitudes of the temperature change $T$ as a function of the optical penetration exponent $n_\alpha$ and the radial wavelength exponent $n_\beta$ at ***(a)*** a low frequency ($0.1\ \omega_r$) and ***(b)*** the resonance frequency ($\omega_r$).

The amplitude surfaces of the temperature change ($T$) for the low, resonance and high modulation frequencies are presented in Fig. 3.a and b. The surfaces seems to consists of two parts; in $n_\beta$ [0,3], a saddle-like surface is dominant, beyond this surface a singularity region is visible. The location of the end of the saddle-like surface as well as the maximum of $T$ in this domain are greatly affected by the change of modulation frequency. At the low frequency (Fig. 3.a), the center of the saddle is around $n_\alpha=1.1$ and it ends around $n_\beta=0$. As the excitation frequency increases, the center moves closer to $n_\alpha=1.8$ and $n_\alpha=2.5$ for the resonance and high modulation frequencies (Fig. 3.a and b, respectively). The end points shifts from $n_\beta=-1/2$ to $n_\beta=-1$ for the resonance and high modulation frequencies. It is noteworthy that the magnitude of the amplitude surfaces substantially decreases as the frequency increases.

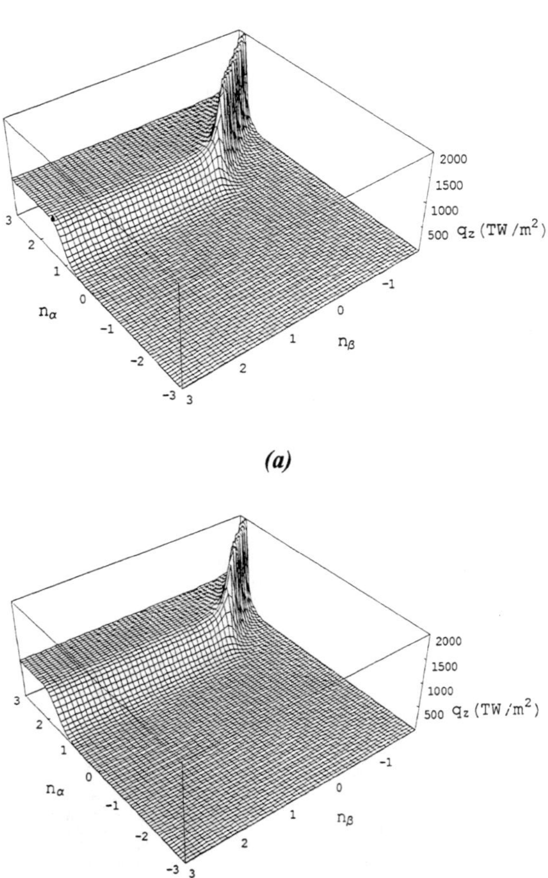


***Figure 4.*** The amplitudes of the heat flux $q_x$ as a function of the optical penetration exponent $n_\alpha$ and the radial wavelength exponent $n_\beta$ at ***(a)*** a low frequency ($0.1\ \omega_r$) and ***(b)*** the resonance frequency ($\omega_r$).

As should be expected, the heat flux in the shallow optical penetration zone approaches to the absorption factors corresponding to skin depths at those excitation frequencies (Fig. 4). The values of the absorption coefficients at the low, resonance and high modulation frequencies are 0.56, 1.06, and 1.56, respectively. As in the temperature case, singularities in the short wave domain appear. The singularity begins approximately at $n_\beta=0$ for the low modulation frequency plot, and $n_\beta=-1/2$ to $n_\beta=-1$ for the resonance and high modulation frequencies, respectively (Fig. 4.a and b). After these shallow penetration limits, the effect of the laser heating is very similar that of the non-penetrating beam.

## CONCLUSIONS AND REMARKS

A major shortcoming encountered in laser ultrasonics is the directionality of the generated wavefield due to the localized nature of the thermal skin. As a result, the longitudinal component of the generated field is weak compared to the shear component. Various solutions to this problem have been introduced, such as operating in the ablation regime and depositing thin liquid films on the substrates. The effect of optical penetration on the wave field has been observed experimentally in the form of a precursor in the waveforms. In this work, a first principle-based fully-coupled thermoelastic formulation based on the equation of motion and heat equation is developed to study the effects of optical penetration in axisymmetric wave propagation in thermoelastic layers and/or layered structures. It is demonstrated that while the optical penetration has no effect on the entries of the sextic transfer matrix, it introduces an equivalent forcing term for all state variables (the displacement vector, stress components, heat and temperature) for both surfaces of a thermoelastic layer. This is opposed to the surface heating case in which the heating effect is localized in the heating volume known as the thermal skin. The thickness of thermal skin depends on the modulation frequency while the optical penetration typically depends on the wavelength of the light alone. This additional forcing vector is functions of the light intensity modulation frequency, the radial wave number, penetration decay rate, as well as thermoelastic material properties. Complexities in wavefields due to the nature of the forcing term are demonstrated.

The magnitudes of the entries of the equivalent forcing vector have been studied in the space of the optical penetration exponent $n_\alpha$ and the radial wavelength exponent $n_\beta$ for three different modulation frequencies. The effects of these factors on the values of the amplitudes and the singularity behavior are studied. From the generated longitudinal stress field data, it is concluded that strong normal stress can be generated in the certain regions of the modulation frequency, $n_\alpha$, and $n_\beta$ space.

The near-field wave fields can be complicated especially for non-metallic materials (*e.g.* polymers, ceramics, and silica). A good understanding of the near-field is essential in thin film and coating applications due to the length-scale effect of the optical penetration. The formulation developed can also be used for layered structures consisting of an arbitrary number of layers. Potential applications of the technique include in *(1)* NDE of thin films and/or layered coatings, *(2)* measurement of the optical penetration depths, (3) use as a semiconductor manufacturing method for opening up interfaces between a film and substrate, and *(4)* study of wavefield and beam directionality due to the optical penetration.

## REFERENCES

1. C.B. Scruby and L.E. Drain 1990 Laser Ultrasonics: Techniques and Applications. Adam Hilger Pub.

2. C. Cetinkaya, Cunli Wu, and Chen Li 2000 Journal of Sound and Vibration 231(1), 195-217. Laser-Based Transient Surface Acceleration for Thermoelastic Layers.

3. C. Sve and J. Miklowitz 1973 Journal of Applied Mechanics, 161-167. Thermally Induced Stress Waves in an Elastic Layer.

4. F. A. McDonalds 1990 Applied Physics Letter56, 230-232. On the Precursor in laser-generated Ultrasound Waveforms in Metals.

5. K.L. Telschow, and R.J. Conant 1990 Journal of Acoustical Soc. of America 88(3). Optical and Thermal Parameter Effects on Laser-Generated Ultrasound.

6. F. Enguehard, L. Bertrand 1997 J. Applied Physics82(4). Effects of Optical Penetration and Laser Pulse Duration on Laser Longitudinal Acoustic Waves.

7. F. Enguehard, L. Bertrand 1998 J of Acoustical Society of America103 (2). Temporal Deconvolution of Laser-generated Longitudinal Acoustic Waves for Optical Characterization and Precise Longitudinal Acoustic Velocity Evaluation.

8. H. W. Lord and Y. Shulman 1967 Journal of the Mechanics and Physics of Solids15, 299-309. A Generalized Dynamical Theory of Thermoelasticity.

9. C. Cetinkaya J. Wu and Chen Li 2000 Journal of Nondestructive Evaluation (in review for publication). An Efficiency Study of Transient Wave Generation in a Thermoelastic Layer with a Pulsed Laser.

10. J. B. Spicer 1991 Ph.D. Dissertation, Johns Hopkins University. Laser Ultrasonics in Finite Structures: Comprehensive Modeling with Supporting Experiments.

11. C. Cetinkaya, A. F. Vakakis, and M. El-Raheb 1995 Journal of Sound and Vibration182(2), 283-302. Axisymmetric Waves in Weakly Coupled Layered Media of Infinite Extent.

12. W. T. Thomson 1950 Journal of Applied Physics 21, 89-93. Transmission of Elastic Waves through a Stratified Solid Medium.

13. N. A. Haskell 1953 Bulletin of the Seismological Society of America 54, 17-34. The Dispersion of Surface Waves on Multilayered Media.

14. C. Cetinkaya and C. Li 2000 the ASME Journal for Vibration and Acoustics 122(3) 263-271. Propagation and Localization of Longitudinal Thermoelastic Waves in Layered Structures.

Proceedings of
2001 ASME International Mechanical Engineering Congress and Exposition
November 11-16, 2001, New York, NY
NDE-Vol. 21

IMECE2001/NDE-25809

# IR THERMOGRAPHY SCANNING TECHNIQUE IN THE ASSESSMENT OF AIRPORT PAVEMENTS

**Antonia Moropoulou**

**Nicolas P. Avdelidis, Maria Koui**

National Technical University of Athens, Department of Chemical Engineering,
Section of Materials Science and Engineering, Iroon Polytechniou 9,
15780 Zografou, Athens, Greece.

## ABSTRACT

Corrosion of asphalt pavements is a major problem involved with materials engineering and transportation engineering. Asphalt pavements frequently start to deteriorate slowly primarily, gradually progressing to failure. Such failure can be expensive in terms of both money and lives, but can also be prevented. For that reason, intelligent techniques should be suggested, with the intention of examining the condition of such engineering structures. IR thermography is a non-destructive technique that has the ability to inspect effectively substantial areas, such as airport pavements. It has been used successfully in the detection of cracks, voids and other imperfections appearing either from the aging of the materials or due to poor workmanship. Investigation of asphalt overlays by the means of IR thermography is possible, given that subsurface defects in a material influence the heat flow through that material, generating surface temperature alterations. Despite that, other issues such as, environmental conditions and surface temperatures should also be considered, in an attempt to obtain reliable results. In this research work, a selection of asphalt pavements located at the International Airport of Athens, in Greece, are investigated by the use of scanning IR thermography. In conclusion, this paper describes the problem of deteriorated airport pavements, the process and the apparatus used for the in situ tests, whilst the results obtained lead to the suggestion of a predictive monitoring non-destructive technique for the inspection and appropriateness of airport pavements.

## INTRODUCTION

Thermography directly senses the infrared radiation that a material or structure emits and detects surface temperature differences [1]. Examination of airport pavements using infrared thermography is feasible, since subsurface defects in a material amend the heat flow through that material, triggering surface temperature variations [2]. Nonetheless, additional aspects such as, environmental conditions, emissivity values and surface temperatures ought to be considered, with the aim of obtaining reliable results [3].

In this research work, various types of airport pavements, situated at the International Airport of Athens in Greece, are investigated by the means of infrared thermography.

## EXPERIMENTAL PROCEDURES & TECHNIQUES

For the inspection of the airport pavements, passive infrared thermographic approach, using the Avio Tvs 2000 Mk II LW, 8-12 µm system, was used. The investigation was carried out through daytime, at times of rapid heat transfer (approximately 3 hours after sunrise), in the course of January, February and March. The surface temperatures of the examined pavements, as well as the environmental conditions, were taken into consideration. The airport pavements were cleaned of all debris and the thermographic device was adjusted to record thermal images of the various pavements studied. Traffic control was also considered, in order to prevent any likely accidents, as well as to keep vehicles away from stopping or standing on the pavements to be tested.

During the scanning procedure of the pavements, the thermographic system was mounted on a vehicle, in accordance to the ASTM standard D4788 – 88 [4], aiming to a fast and accurate investigation. At the same time a laser type range finder (LTI 20-20 Ultralyte) was also used, in order to determine the boundaries of the investigated sections, as well as the identified defected areas that were spotted during the infrared thermographic scanning procedure.

Finally, it is worth mentioning that the defects were identified in accordance to the ASTM standard D4788 – 88, where there must be temperature differences of at least 0.5 Celsius degrees, between the defected areas and the solid pavements.

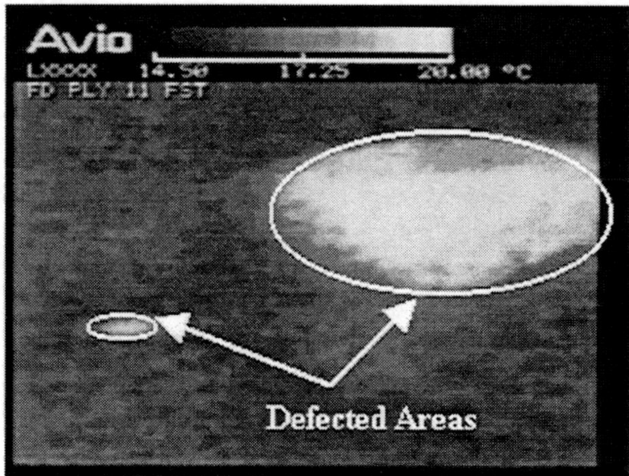


**Figure 1: Defected areas within the 321 section**

## RESULTS & DISCUSSION

Sections from three different pavements within the airport were inspected by the use of infrared thermography. The first section examined, was the northwest part of the taxiway (section 321), which is actually in service, in order to accommodate, small private planes and helicopters. The dimensions of this investigated section were: length of 425 m and width of 45 m. For this section of the airport, crushed rock is used as a paving material. This is one of the oldest sections within the airport and although it is in an up to standard state, it presents aging signs. Secondly, the bravo taxiway was investigated. The dimensions of this investigated section were: length of 350 m and width of 45 m. This pavement is consisted of a brand new asphalt overlay. Finally, the main runway of the airport was also inspected. This pavement is also consisted of a new asphalt overlay of approximately 30mm thickness, recently replaced. The dimensions of the sections investigated within the main runway were: length of 746 m and width of 56 m.

Representative results from the three investigated pavements are presented.

In figure 1, surface voids as a result of materials aging, arising from the investigation of the northwest part of the airport (section 321), are presented. These surface voids are limned as yellow – "hot" areas. These two yellow colored spots present the greater temperature differences, i.e. much warmer than the remaining temperatures within the image.

In figure 2, an investigated section from the northwest part of the taxiway (section 321) is presented. Once again, there is a scanned defected area (limned as yellow – "hot" area) presented, as a result of materials aging.

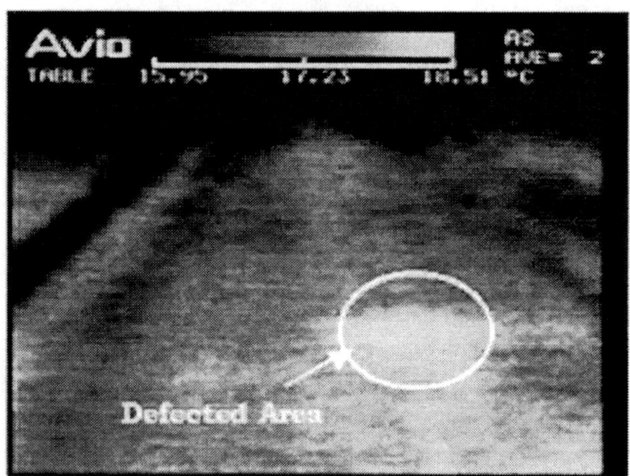


**Figure 2: Defected area from section 321**

In figure 3, a crack is located, due to the temperature variation when compared to the rest of the image. Since the examined pavement was recently replaced, this localized crack is attributed to the poor workmanship on this specific examined section of the Bravo Taxiway.

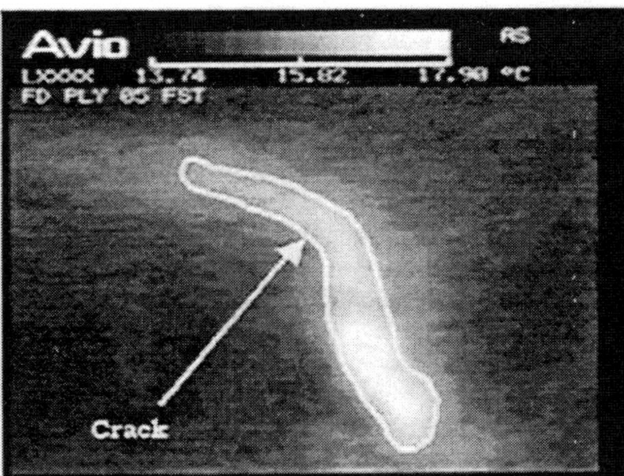


**Figure 3: Location of crack on the Bravo Taxiway**

In figure 4, two fundamental imperfections are presented. Once again, these defects are visualized with yellow – "hot" colors, representing approximately the 10% of the thermograph. This percentage designates the defectively section of the investigated pavement.

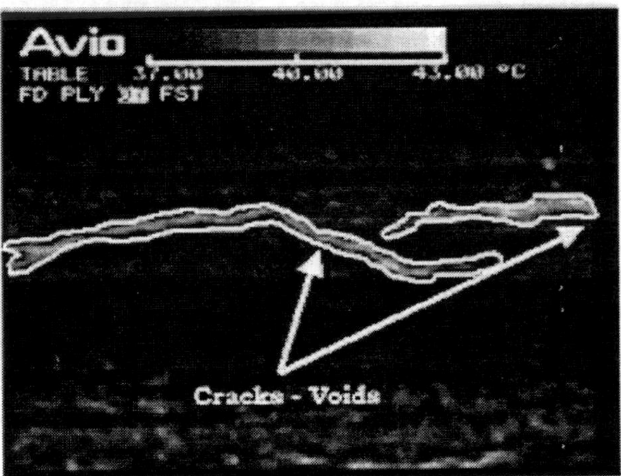


**Figure 4: Cracks – Voids at the main runway**

Although, the thermograph in figure 5 shows a recently replaced pavement, it presents an imperfection on the surface of the main runway. The significant temperature dissimilarity of this imperfection exhibits the importance of the deteriorated area of the investigated pavement. This imperfection occupies approximately the 12% of the above-mentioned thermograph.

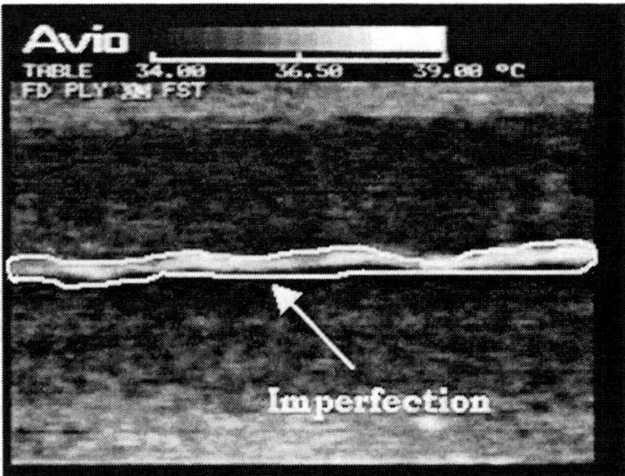


**Figure 5: Imperfection at the main runway**

At this point, it is essential to reveal a number of aspects that were taken into account, in order to obtain accurate and reliable results.

- Cloud cover: The survey was carried out at periods of unimportant or no cloud cover. This endorsed the most effective energy transfer between the sky and the investigated surfaces.
- Ground moisture: Since moisture affects the surface temperature and therefore the imperfections, the investigation was completed at times where the surfaces were dried out.
- Solar radiation: The inspection was carried out during daylight, whereas the occurrence of solar radiation produces largely rapid heat transfer.
- Surface texture: Since the surface emissivities are affected by surface dissimilarities, minor adjustments were made in the focus, in the contrast and brightness controls.

All the above-mentioned parameters should at all times be taken into account, so that the investigation would be accurate and in accordance to the ASTM standard D4788 – 88. Finally, taking into consideration that the investigation was completed during mild environmental conditions, the defects arising from this investigation should be significantly considered.

The results obtained from the non-destructive investigation provide a prompt and accurate assessment tool of the examined pavements. Specifically, the northwest part of the taxiway has not been repaired or condition assessed, at least not in the last decade, mostly due to the fact that is in service in order to accommodate small private planes and helicopters. In addition, the Bravo Taxiway and the main runway of the airport, which undergone a recent repair (replacement of the asphalt overlay), have been inspected at various times. Those inspections included sampling collection from random pavement areas, in order to test out their strength [5]. The results from those mechanical tests confirm in a way the infrared thermographic results. Nonetheless, as the International Airport of Athens is out of use since April 2001 and a new Airport is currently used (El. Venizelos Airport), further repairs will not be made, from the proofs arising from this research work.

## CONCLUSIONS

Detection of cracks, surface voids and other anomalies on asphalt overlays, is possible by applying infrared thermographic techniques. The advantage of thermography over the destructive testing methods is that large areas, such as airport pavements, can be scanned fast and with no need to be destroyed throughout testing. This results in major savings in time, people, work and equipment. Also, no material remains are formed which could trigger environmental problems. There are also advantages of infrared thermography over other non-destructive techniques. The equipment that is intended for infrared thermographic investigations is safe, as it does not emit any radiation. It is only used for recording the thermal radiation emitted from the material – surface under examination. In addition, infrared thermography is an area investigating technique, while most of the other non-destructive methods are either point or line testing methods. Furthermore, infrared thermographic testing may be performed during both day – and night – time hours, depending upon the environmental conditions.

The main limitation of the infrared thermographic technique is the lack to determine the depth and thickness of cracks or voids. Nonetheless, by using other non-destructive techniques, such as ground penetrating radar, it is possible to obtain information concerning the depth and thickness of cracks, voids or other anomalies appearing on materials [6].

Conclusively, the sustained evidence implied by this work, results to the suggestion of a monitoring scanning non-destructive technique, capable of assessing the efficiency of airport pavements.

## ACKNOWLEDGEMENTS

Acknowledgements are attributed to the Hellenic Civil Airports Authority for the assignment of this pilot investigation program to the National Technical University of Athens, from which the results of this work arise. The authors would also like to thank E.T. Delegou, N. Gousios, G. Haralampopoulos, K. Kakaras, M. Karoglou, D. Lountzis, K. Roumpopoulos and I. Tzevelekos, for their assistance in the pilot investigation of asphalt pavements at the International Airport of Athens, Greece.

## REFERENCES

1. Moropoulou, A., Koui, M., Avdelidis, N. P., Kakaras, K., 2000, "Intelligent passive control methods to inspect quality of airport runways and asphalt pavements," Proceedings, Computational Methods for Smart Structures and Materials Conference, C. A. Brebbia, A. Samartin, eds., WIT Press, pp. 85-93.
2. Moropoulou, A., Koui, M., Avdelidis, N. P., Kakaras, K., 2000, "Inspection of airport runways and asphalt pavements using long wave infrared thermography," Proceedings, Thermosense XXII Conference, R. B. Dinwiddie, D. H. LeMieux, eds., SPIE Press, Florida, Vol. 4020, pp. 302-309.
3. Moropoulou, A., Avdelidis, N. P., Koui, M., 2000, "Detection of defects of airport pavements using infrared thermography," Journal of the Canadian Society of NDT, **21(5)**, pp. 5-8.
4. ASTM D4788-88. Test method for detecting delaminations in bridge decks using infrared thermography.
5. Abakoumkin, K., Loizos, A., Papavasileiou, V., Xaronitis, G., 1998, "Development of a system for airport pavements assessment and maintenance", Technical Report, National Technical University of Athens, Greece.
6. Moropoulou, A., Avdelidis, N. P., Koui, M., Aggelopoulos, A., Karmis, P., 2001, "Infrared thermography and ground penetrating radar for airport pavements assessment", Journal of Nondestructive Testing & Evaluation, in press.

Proceedings of
2001 ASME International Mechanical Engineering Congress and Exposition
November 11–16, 2001, New York, NY
NDE-Vol. 21

IMECE2001/NDE-258010

# TRENDS IN NDE USING THERMAL INSPECTION METHODS

Norbert G. Meyendorf

Jochen Hoffmann, Liming Shen, Henrik Rösner*

Center for Materials Diagnostics, University of Dayton, 300 College Park, Dayton, OH 45469-0121
*Fraunhofer Institute (IZFP), Universität, Gebäude 37, D-66123 Saarbrücken, Germany

## ABSTRACT

This paper gives a brief overview of present NDE techniques that use infrared cameras. Starting with physical basics, current state of the art and future trends are discussed. Two new methods for materials characterization that use heat dissipation and hot air for heat stimulation will be presented. A new technique called fan thermography, has high potential for corrosion detection under intact coatings. Using this technique, low-cost uncooled infrared cameras are sufficient for obtaining excellent, highly sensitive imaging results.

## 1. INTRODUCTION

NDE for the 21st century means testing and control techniques integrated in fully automated production processes. Fast, no contact and real-time techniques will be necessary to meet these requirements. Testing of high-integrated micro/nano-structures will require both high spatial and temperature resolution of these techniques. Test methods based on optical or infrared imaging are favored to do this job. While optical techniques are able to detect surface defects and surface structures, thermographic methods are sensitive to invisible and subsurface defects. Uncooled and cost efficient micro bolometer cameras are now available that have very low requirements for service and maintenance. Thus, thermal techniques become interesting for a wide variety of NDE applications. Thermography has also high potential for condition monitoring. This paper will discuss possible applications for early detection of corrosion and fatigue for aging aircraft.

## 2. PHYSICAL BACKGROUND

Thermal imaging techniques are based on the fact that every object emits electromagnetic waves. Spectral distribution and intensity of this radiation depend on the surface temperature. The spectral radiation density can be calculated using the Plank's radiation, which is valid for an object with the highest possible emissivity (the black body). The spectral radiation density L of a real object is achieved by multiplication with an emissivity coefficient ε (ε<1).

$$L(\lambda,T)=\varepsilon\frac{2hc^2}{\lambda^5}\left[e^{\frac{hc}{\lambda kT}}-1\right]^{-1}$$

| | | | |
|---|---|---|---|
| L(λ,T): | Spectral radiation density | | |
| ε: | Emissivity | c: | speed of light |
| λ: | Wavelength | T: | absolute temperature |
| h: | Planck's constant | | |

The wavelength dependent emissivity coefficient ε(λ) describes the fraction of emitted radiation intensity from the real objects compared to the black body and is always smaller than 1. Unfortunately, ε becomes very small for metals in the infrared range (0.1 or even less for aluminum surfaces at wavelength 10 μm). The sum of emissivity (ε), transmissivity (τ) and reflectivity (ρ) has to be 1 for a selected material (ε+τ+ρ=1). Therefore most metal surfaces are excellent mirrors for infrared radiation. For thermographic NDE of metallic parts, it is usually required to apply a black coating to their surfaces, which has to be removed after testing. For testing of ceramics or plastics, it has to be considered that these materials might be transparent or opaque to infrared radiation.

These spectral properties have to be considered if an infrared camera type is selected for a certain test problem. Usually, cameras for two wavelength ranges are available, short wave cameras for the wavelength range 3 to 5 μm and long wave cameras for 8 to 12μm range. These wavelength ranges are the "atmospheric windows" where air has the highest transparency. Molecular gases absorb and emit infrared radiation at a certain wavelength above 1 μm. Within the atmospheric windows are no major absorption band for gases that form clean air. However, this can be different in industrial atmospheres, where effects of gasses have to be considered for NDE.

## 3. THERMAL CAMERAS FOR NDE

Older Infrared cameras are scanner systems. They have only one infrared detector. Frames are generated by scanning across the surface (usually by means of mirror systems). It has to be considered for NDE tasks with high temperature dynamics, such as testing of thin coatings or metallic objects, that each pixel represents another testing time for scanner cameras. Focal Plane Array (FPA) cameras that make use of a multi-sensor array are available with different sensor types, different pixel sizes and different frame rates (see table 1).

For most NDE applications, cameras with excellent temperature resolution are necessary. Testing procedures require high temperature dynamics, if very thin or high thermal conducting films have to be tested. Therefore, frame rates higher than video standard (USA: 60 half frames per second) are necessary. Cameras based on Quantum Well Infrared Photon detectors (QWIP) are available with excellent temperature resolution (Noise equivalent temperature difference of a few mK) and high pixel resolution. Some of these high-end cameras generate frame rates of up to 7000 per second. Unfortunately, the high price of these cameras is a limiting factor for a wide variety of applications.

**Table 1. Cameras Available for Thermal NDE Techniques**

| **Detector Material** | **Si** | **Micro bolometer** | **HgCdTe** | **PtSi** | **HgCdTe** | **QWIP** |
|---|---|---|---|---|---|---|
| **Detector Type** | **CDD array** | **Array** | **Scanner** | **FPA** | **FPA** | **FPA** |
| **Noise Equivalent** | | **80 mK** | **150 mK** | **75 mK** | **25 mK** | **down to 7mK** |
| **Lowest Detectable Temperature** | **400°C** | **-20 °C** | **-30 °C** | **-30 °C** | **-30 °C** | **-50 °C** |

Micro bolometer cameras could be interesting for a lot of NDE applications. An important advantage of these systems is that the detector does not require coolant like liquid nitrogen or a mechanical cooling system (Sterling motor) with limited lifetime. Thus, these systems are low priced and have low service and maintenance requirements. Unfortunately, these cameras have poor temperature sensitivity (high noise) compared to QWIP systems.

It is possible to compensate this disadvantage by averaging of a high number of video frames during one test. Assuming white noise, an average of 100 frames will increase the temperature resolution by a factor of 10. Averaging of a high number of frames requires NDE test procedures with low temperature dynamics, different to the presently used flash or lock-in techniques. One example will be given below.

Nevertheless, even conventional CCD cameras based on silicon detectors can be used for several applications. These cameras are available for a few hundred dollars. If filters for infrared radiation are removed from the CCD-chip near infrared radiation from 0.8 to 1.1 μm wavelength is detected. This is sufficient for temperatures above 400°C. The author has successfully applied CCD cameras to control of welding processes [1].

## 4. THERMAL NDE METHODS

Passive thermal NDE techniques detect heat radiation emitted from a test object without external heating. An example is the control of welding processes. An interesting application is the inspection of internal structures in silicon-microchips. Silicon is transparent for infrared radiation. Therefore, infrared cameras can image internal structures of the chip [2].

Most thermal NDE techniques are based on active principles. The test object is stimulated by external heat sources, for example light or thermal radiation, electric current, microwaves, or mechanical loads. The thermal camera detects the surface temperature variations, due to these effects. The thermoelastic effect described by Joule Thomson Law is responsible for temperature oscillations ($\Delta T_{el}$) if a cyclic load ($\sigma_a$) is applied:

$$\Delta T_{el} = \frac{-\alpha \cdot T_0 \cdot \sigma_a}{\rho \cdot C_p}$$

Here, $T_0$ is the absolute temperature, $\alpha$ is the coefficient of thermal expansion, $\rho$ is the density and $C_p$ is the specific heat at constant pressure. This principle is used for imaging of local stress concentrations and for detection of critical locations in components where fatigue cracks might appear first [3]. Beside thermoelasticity due to the mechanical hysteresis cyclic loading results in the generation of dissipated heat. By averaging thermal images for a large number of cycles, the temperature enhancement due heat dissipated heat can be separated from thermoelastic effects.

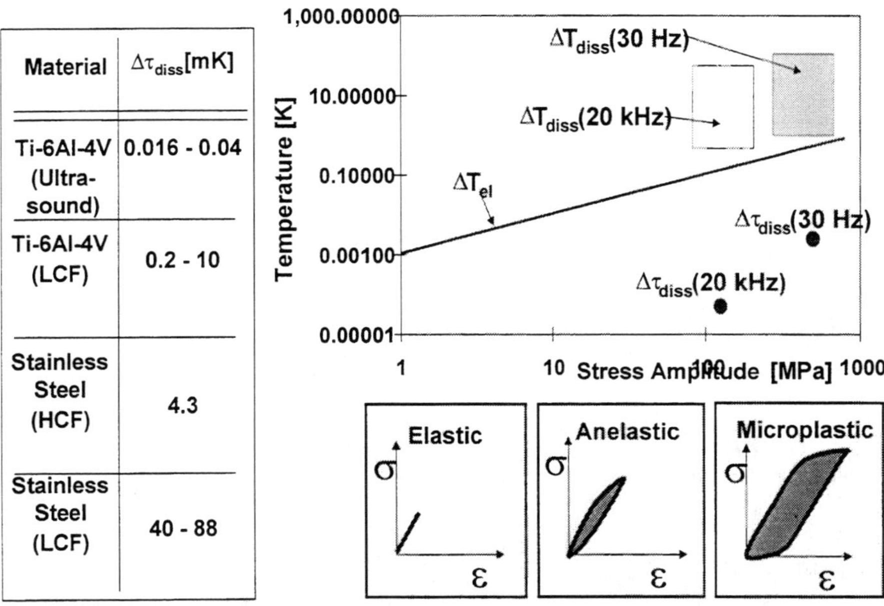

| Material | $\Delta\tau_{diss}$[mK] |
|---|---|
| Ti-6Al-4V (Ultra-sound) | 0.016 - 0.04 |
| Ti-6Al-4V (LCF) | 0.2 - 10 |
| Stainless Steel (HCF) | 4.3 |
| Stainless Steel (LCF) | 40 - 88 |



**Figure 1. Thermoelastic ($\Delta T_{el}$) and dissipative ($\Delta T_{diss}$) heating effect during cyclic loading of titanium Ti-6Al-4V; Temperature increase due to one single loading cycle ($\tau_{diss}$) caused by dissipation.**

It has been demonstrated in [4] that heat dissipation can be used for microstructure characterization of a metal and for prediction of fatigue life. Thermal imaging during stimulation with high power ultrasound is a possible method for fatigue characterization [5].

Defects (cracks, delaminations) cause locally enhanced mechanical losses due to a cyclic mechanical loading. The reason can be the friction of crack's internal surfaces, delaminations or a stress concentration at the defects. Thus, locally significantly higher temperature effects can be generated compared to the thermoelastic or integral heat dissipative temperature effect. Due to this local conversion of the mechanical energy to heat energy on cracked material, defects can be visualized as hot spots in a thermal image. This effect is the basis of Vibro-Thermography using low frequency loading [6], Lose Angel Thermography [7] or Thermosonic Testing using high power ultrasound [8]. Ultrasonic frequencies can be very efficient in heating of technical components due to the high rate of hysteresis cycles in the material.

At present, the most used thermal NDE technique is flash thermography. An energy pulse (usually light) heats the surface of the test object and the cooling process is observed by an infrared camera. Subsurface defects disturb the cooling process and result in a temperature contrast after a certain response time. This method is for instance efficient for characterization of coatings and detection of subsurface defects. Examples are the characterization of corrosion or detection of defects in adhesive bonds [9]. A modification of flash thermography is line-scanning thermography where a heat line is scanned across the surface. The temperature field behind the heat line represents the cooling process [10]. The distance between heat generation and detection determines the response time $t_r = s/v$ where s is generation to detection distance and v is scanning speed.

Lock-in thermography detects the same type of defects like flash thermography. Here the heat is stimulated periodically. This allows a more detailed data analysis and as some authors report a higher depth sensitivity for the phase signal. However, this method requires significantly more testing time [11].

Table 2 summarizes different thermographic techniques. These methods vary with regard to the mechanism of heat generation and the time dependence of the temperature measurement.

**Table 2. Different Thermographic Techniques**

| Temperature Measurement | Physical Mechanisms Used for Heat Generation |
|---|---|
| • During Periodical modulation (Lock-in Thermography<br>• During Cooling phase after an heat pulse (Flash Thermography)<br>• Behind a heating line scanned across the surface (Line Scanning Thermography)<br>• During the heating process (Thermosonics, Fan Thermography) | • Absorption of light<br>• Absorption of infrared radiation<br>• Absorption of microwave radiation<br>• Inductive heating<br>• High power ultrasound<br>• Hot air<br>• Mechanical loading |

## 5. FAN THERMOGRAPHY – A COST EFFECTIVE WAY FOR HIGH-RESOLUTION THERMAL NDE

Disadvantage of most thermal NDE principles is the low temperature contrast as well as high temperature dynamics (if thin layers are tested) that requires high sensitive and high-speed thermal cameras. A model simulation for testing of a polymer coatings using flash thermography illustrates the problem. This simulation is based on the thermal diffusion equation

$$c_p \rho \frac{\partial T}{\partial t} = \lambda \frac{\partial^2 T}{\partial x^2} + I_0(t)\gamma e^{-\gamma x}$$

$C_p$: specific heat; $\lambda$: heat conductivity, $\rho$: density, $\gamma$: absorption coefficient, $I_0(t)$: absorbed radiation

Perfect Adhesion

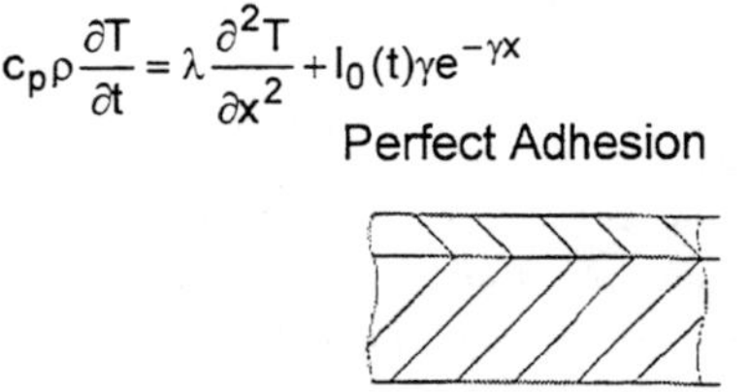

Delamination

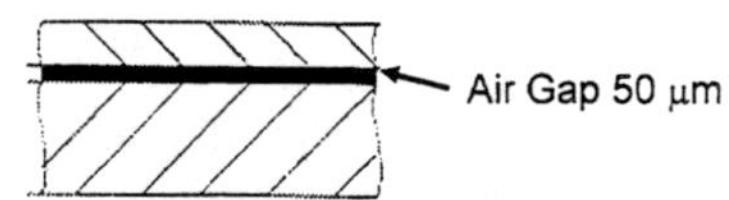


Surface Temperature Enhancement

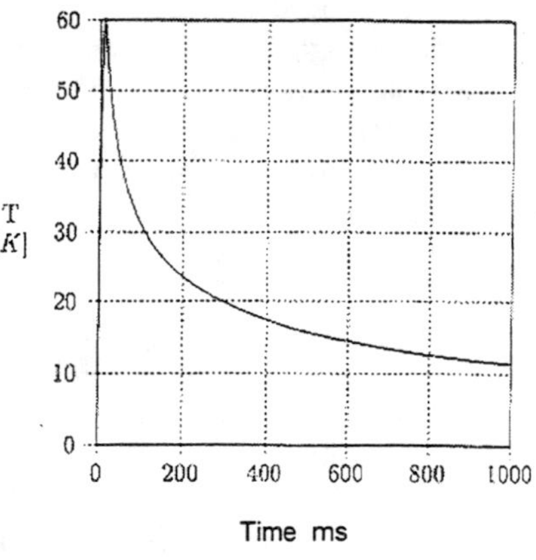


Temperature Contrast

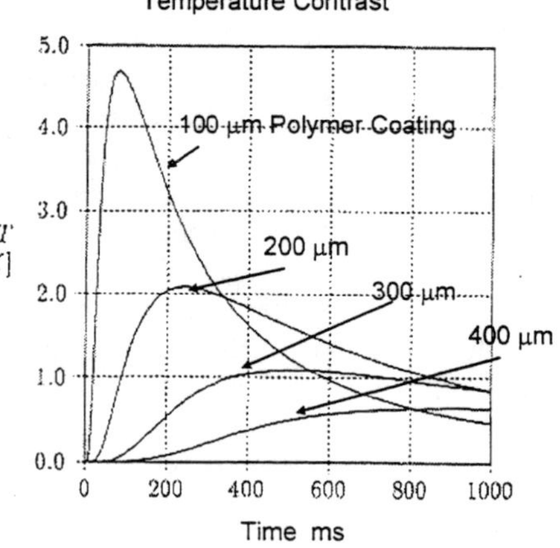


:

The heat generated by a short pulse at the surface is assumed to diffuse into the structure. Layers of a polymer and air (for the assumed delamitation) of different thickness modify the cooling process.

**Figure 2. Simulation of surface temperature for flash thermography testing: Left: Surface temperature/ time diagram; Right: Temperature contrast due to a delamination [12].**

To detect the assumed delamination of a 400 µm coating on aluminum, a temperature dynamics of 50 K is necessary for a temperature contrast of approximately half a degree Kelvin. For thinner coatings the contrast increases, but the response time decreases. For the 100 μm coating, the maximum contrast appears only for the time of one video frame (50ms). Therefore, image averaging to enhance temperature resolution can not be used.

A possible way to avoid this problem is to measure the temperature contrast during a slow heating process or in a quasi-static equilibrium between the absorbed heat energy and the heat energy losses. Unfortunately, radiation can not be used simultaneously for stimulation and detection from the same side of the test object because reflected heating radiation and emitted body radiation are interfering. A very simple application of this principle was based on detecting of surface temperature by liquid crystals [13]. A different way is to use hot air from a fan for heating and (cheap micro-bolometer) infrared camera for detection. Because cameras are sensitive in the atmospheric windows, hot air will not effect the infrared image.

We assume heating and temperature measurement from the same side of the object and approximate the test procedure by quasi-static conditions. This means we assume a constant heat current I=Q/t (Q: heat energy) through the coating. The metal substrate is assumed to remain at a constant temperature. If the surface temperature ($T_{surface}$) enhancement is small compared to the air temperature ($T_{air}$), the heat current is determined by the heat transition from air to coating: $I=Q/t = \alpha * A (T_{air} - T_{surface})$. $\alpha$ is the heat transfer coefficient and is approximately $15 W/m^2K$ for an air velocity of 5m/s [14]. For heat conduction through coating and air gap, the heat conductivity equation $I=Q/t = \Delta T/R$ is valid. $R= d/(\lambda * A)$ is the heat resistance where $\lambda$ is the heat conduction coefficient for the layer material, d is the layer thickness and A is the area where heat conductivity appears. For the assumed test procedure, A is the same for heat transition from air to surface and heat conduction into the test object. Due to the analogy to the Ohm's law a series circuit of (thermal) resistivities can be considered for a multilayer problem such as coating and air gap. From the formulas above the surface temperature enhancement for quasi-static conditions can be calculated for an n-layer system:

$$\Delta T = \alpha * (T_{air} - T_{surface}) [\Sigma(d_n/\lambda_n]$$

For equal geometrical and thermal parameters as assumed for flash thermography in Figure 2 ($\lambda_{coating}$=0.17W/m K; $\lambda_{air}$=0.025W/m K) we get the following results:

**Table 3. Temperature Contrast for Fan Thermography (Air Velocity = 5m/s, $\alpha$=15W/m$^2$K; Air Temperature = 150°C)**

| Coating Thickness | 100μm | 200μm | 300μm | 400μm |
|---|---|---|---|---|
| ΔT without air gap | 1.15 K | 2.30 K | 3.45 K | 4.60 K |
| ΔT with air gap 50μm | 5.05 K | 6.25 K | 7.35 K | 8.50 K |
| Temperature contrast | 3.9 K | 3.9 K | 3.9 K | 3.9 K |

Even for the thinnest coating, the contrast is in the same range as for flash thermography. But for fan thermography we have a stable contrast for a much longer time period. So by image averaging the accuracy can be further improved. Image averaging allows using cheap uncooled bolometer cameras, which have a higher noise. Image averaging reduces this noise drastically. The quasi-static conditions should also allow testing of thin coatings with high thermal conductivity without high-speed cameras. Flash thermography requires high speed because of the short time period for thermal exchange processes in this case.

## 6. EXPERIMENTAL RESULTS

Two examples demonstrate possible applications.

In the first example, corrosion on aluminum has been detected under an intact coating. By measuring an epoxy, 50 μm thick, after 125 hours exposure to corrosive environment, Figure 3 compares Scanning Vibrating Electrode Technique (SVET) with Fan Thermography images of the same sample area. First SVET measurements were carried out. Figure 3a presents a vector overlay mapping of electrochemical current density superimposed over a video image. Delaminations can be observed in the video image. The longer vectors indicate, the higher corrosion activity underneath the coating is.

Coating sites of corrosion or delaminations are thermal barriers. Locations of enhanced corrosions from the SVET image correspond to hot spots in the fan thermographic image (Figure 3b). For this experiment an air stream of 150°C heated the surface of the coating. The thermal image was the result from an averaged sequence of approximately 50 frames.

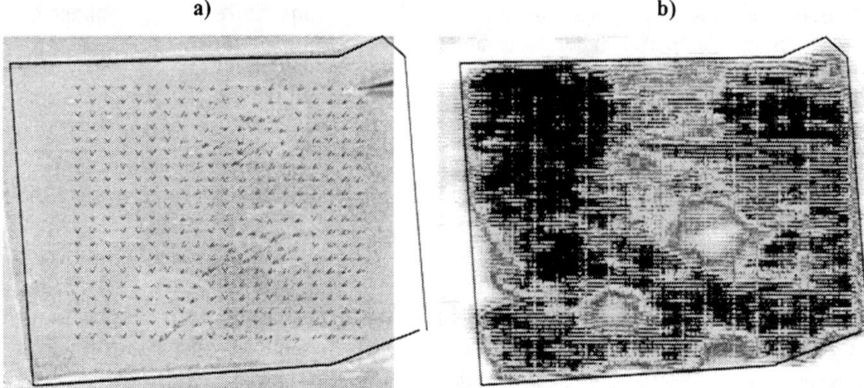


**Figure 3. NDE imaging of corrosion under 50 μm thick artificially damaged epoxy coating**
**a) SVET video image + current density vector overlay (vector area: 4x4 mm$^2$)**
**b) Fan Thermography image (image area: 6x6 mm$^2$).**

The second example shows thermal testing of three brake pads. A brake pad is made by a 3 mm metal substrates with a 6 mm ceramic materials (brake lining) glued on top. The parts where heated from the metal side and the temperature was measured at the lining surface (Figure 4). Cracks in the lining or bonding defects reduce the heat flow through the part. Therefore, defects are indicated by lower temperature (dark) in this case.

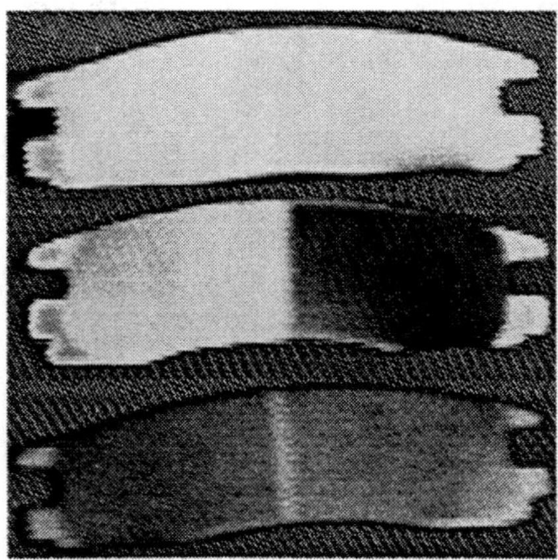

**Figure 4. Fan Thermographic Testing of Brake Pads.**

The upper pad was defect free. In the pad below, a large crack in the right sector is clearly visible. In the third pad (bottom) a slightly lower temperature at the right sector indicates a leak of adhesion, not visible from outside. To enhance the visible effect, the grayscale was modified by a constant offset compared to the other frames.

## 7. CONCLUSIONS

New thermal imaging techniques will push thermographic NDE methods. Because these techniques are fast, non-contact, and full field techniques they are suitable for a wide variety of applications and are the favored candidates for NDE in the 21st century. Thermographic NDE techniques mainly differ in the way of stimulation of a thermal signal and in the analysis of thermal frame sequences. The best chance for wide varieties of applications have those techniques, which use cheap and uncooled cameras with long lifetime and low service requirements. Fan thermography and acousto-thermal methods are possible candidates, because they allow averaging of frame sequences for data analysis.

## 8. REFERENCES


1. Meyendorf, N., Ehrlich, S., and Nitzsche, R., "Thermografische Ueberwachung von Schweissprozessen", Bild und Ton 45, 3/4 (1992), pp. 55.
2. Walle, G., Karpen, W., Banjan, P., Becker, D., Rösner, H., Meyendorf, N., and Netzelmann, U., "Schichtprüfung mit Hochgeschwindigkeits-Thermographiekameras, Modellierung und Werkstoffcharakterisierung" FhG-IZFP Report 010111-TW (2001).
3. Harwood, N., and Cummings, W. M., "Thermoelastic Stress Analysis", p. 35-43, Adam Hilger, Bristol, (1991).
4. Rösner, H., Meyendorf, N., Sathish, S., and Matikas, T.E., "Nondestructive Evaluation of Fatigue in Titanium Alloys", MRS Symp. Proc. Vol. 591 (2000), pp. 73.
5. Rösner, H., Sathish, S., and Meyendorf, N., "Thermographic Characterization of Fatigue", Rev. of Progress in QNDE, Ames, Iowa 2000, Vol. 20B, pp.1702.
6. Fischer, G.: "Vibrothermographie – ein zerstörungsfreies Verfahren zur Verformungsanalyse und Schadensdetektion bei Kunststoffteilen. VDI Berichte 631, VDI Verlag Düsseldorf, (1987), p. 153-162.
7. Rantala, J., Wu, D., Salerno, A., and Busse, G.: Lockin-thermography with mechanical loss angle heating at ultrasonic frequencies. Busse, G.; Balageas, D.; and Carlomagno, G.M. (Hrsg.), Quantitative infrared thermography, QIRT 96. Pisa: Edizione ETS, (1997), pp. 389-393, ISBN 88 - 467 - 0089 – 9.
8. Favro, L.D., Han, X., Li, L., Zhong, O., Sun, G., Thomas, R.L., and Richars, A., "Thermosonic Imaging for NDE"; Rev. of Progress In QNDE, Ames, Iowa 2000, Vol 20A, pp. 478.
9. Hoffmann, J.P., Matikas, T.E., Sathish, S., Khobaib, M., Meyendorf, N., and Netzelmann, U., "Nondestructive Characterization of Organic Corrosion Protective Coatings on Aluminum Alloy Substrates", 3rd Annual Report for DARPA-MURI, Grant Number F49620-96-1-0442, Dayton, OH (1999).
10. Walle, G.; Karpen, W., Netzelmann, U., Rösner, H., and Meyendorf, N., "Nondestructive Testing with Thermographic Techniques. In: Technisches Messen", ATM, TM 66 (1999), Nr.9, S. 312-321.
11. Busse, G., Wu, D., and Karpen, W., "Thermal wave imaging with phase sensitive modulated thermography", J. Appl. Phys. Vol. 71 (1992), pp. 3962-3965.
12. Burgschweiger, J., "Simulation von Temperaturfeldern bei der zerstörungsfreien Werkstoffprüfung mittels der Impuls-Video-Thermographie" Master Thesis, University Magdeburg (1993).
13. Moore, P.O., and McIntire, P., Nondestructive Testing Handbook, Ed.2, Vol. 9, Special Nondestructive Testing Methods, ASNT (1995).
14. Mende, D., and Simon, G., "Physik Gleichungen und Tabellen", Fachbuchverlag, Leipzig (1971).

**Proceedings of**
**2001 ASME International Mechanical Engineering Congress and Exposition**
**November 11–16, 2001, New York, NY**
**NDE-Vol. 21**

# IMECE2001/NDE-25811

## Development of a Health Monitoring Capability for FRP using Optical Fiber Sensors

Christopher Cassino, ESM VT
John Duke, PhD. ESM VT
Jason Borinski, Luna Innovations

Engineering Science and Mechanics Department
Mail Code 0219, Norris Hall
Virginia Polytechnic Institute and State University
Blacksburg VA 24061
(540) 231-4870, ccassino@vt.edu

**ABSTRACT**

*Fiber reinforced polymers (FRP) are an efficient and relatively inexpensive method of repairing or replacing deteriorating and damaged infrastructure. FRP sheets can be applied to spalling bridge sections to prevent further deterioration and increase stiffness. Pre-fabricated FRP decks can also be used to replace weak portions of the bridge superstructure. However, the effect of the environment on long-term durability of FRP and how the various mechanisms of deterioration initiate and develop are not known.*

*Systems for health monitoring are being sought as a means of managing transportation structures as assets in light of modern life cycle cost concepts. This paper will describe the efforts to develop a method to monitor the structural integrity of FRP materials utilizing an optical fiber sensor developed by Luna Innovations. This sensor is capable of detecting both transient, acoustic emission strain events as well as the accumulated strain in the components to which it is attached. The fact that this sensor is impervious to electromagnetic radiation makes it ideal for monitoring FRP in transportation applications.*

Keywords: Fiber Reinforced Plastics, Nondestructive Evaluation of Infrastructure, Acoustic Emission, Structural Health Monitoring, optical fiber sensors, EFPI

## 1 Introduction

Almost one-third of the 581,862 bridges over 6 meters in length in the United States of America are structurally or functionally deficient according to the standards set forth by the American Society of Civil Engineers [1]. A functionally deteriorated bridge cannot support its designed service levels of traffic load and speed. A structurally deficient bridge is a hazard and must be repaired. High strength composites represent an ideal way to repair bridge structures. Sheets of unidirectional fiber reinforced plastics (FRP) can be applied to the underside of bridge decks to reinforce aging structures and increase stiffness and allowable loads. Layers of FRP can be wrapped around deteriorated columns and beams to strengthen and provide exterior confinement. However, little is known about the long-term durability of composites exposed to the environment during their service life. A potential method of evaluating the long-term structural health of FRP in infrastructure applications is monitoring the acoustic emissions (AE) from the composite.

A fiberglass composite box beam was instrumented with sensors to determine the effectiveness of monitoring the AE resulting from loading and other sources. A series of extrinsic Fabry-Perot interferometer (EFPI) AE sensors, based on the concept developed by Murphy et al. and developed by Luna Innovations, were used to detect AE from pencil lead breaks (PLB) and loading. The EFPI response was calibrated by comparing the AE from a PLB source on an Aluminum plate to a NIST conical transducer and a Dynamic Finite Element Model (DFEM) developed by Prosser [2]. The response of the EFPI on the fiberglass structure was compared to a Physical Acoustics WD Acoustic Emissions sensor.

## 2 Acoustic Emission Detection and Monitoring

Acoustic emissions result from the spontaneous release of mechanical energy that occurs whenever a material deforms, or goes through an irreversible deformation process. Events such as cracking, delamination, fiber failure and chemical reactions all cause the sudden release of mechanical energy [3]. It is important to note that the release of energy is spontaneous, but not instantaneous. Since most engineering materials are deformable, the energy is released by means of a stress wave propagating out from the point in the body that generated the AE. By determining the source location and the components of displacement at the receiver location it is possible to compare the experimentally detected response to that predicted for a hypothetical source disturbance.

Various criteria are used to evaluate AE events. The maximum recorded amplitude of the event, the number of events above a threshold value, the frequency of the AE signals or the total energy measured from the waveform are all used to describe the event and decide what to do based on the results [4]. The results from AE monitoring are often used to isolate regions for further testing by other non-destructive methods, such as ultrasonic testing.

Piezoelectric transducers are commonly used as AE sensors. These transducers are most sensitive to the component of material displacement normal to the surface. However, these transducers are sensitive to both in-plane and out-of-plane displacement thus making it impossible to distinguish the direction of the in-plane component of the displacement experienced by the sensor associated with the event. Piezoelectric transducers are also sensitive to electromagnetic interference that can be present in the field.

The characteristics of piezoelectric transducers are not ideally suited for monitoring AE in fiber reinforced composites. Often, the main AE source of interest in FRP is fiber rupture along a main axis of reinforcement. Since the reinforcing fibers act as a wave guide, the component of displacement normal to the fiber axis may not be as significant as the component in the direction of the fiber. Vehicle generated electromagnetic interference is also a concern when monitoring infrastructure because it is often inconvenient and expensive to block or reroute traffic access to the structure.

## 3 The Extrinsic Fabry-Perot Interferometer

The EFPI is different from a conventional piezoelectric sensor in two important aspects. First, the EFPI is capable of detecting only the in-plane component of displacements. Second, it uses light and the photo refractive effect to determine the magnitude of the displacement detected by the sensor. A resonant interferometric cavity is created by aligning two single mode optical fibers with silvered ends that reflect and transmit the light from a laser source. As a stress wave with passes through the material that the sensor is bonded to, the physical displacement causes the gap between the two fibers to change, which alters the phase and amplitude of the reflected light. The sensor is attached to the surface with epoxy at points equidistant from the center of the cavity. A fast cure epoxy (90 seconds or less) is typically used for installation. The gage length of the sensor is the distance between the epoxy attachment points, typically 10-15 mm.

The EFPI sensor is ideally suited for monitoring AE in FRP composite materials for several reasons. Because the sensor is uniquely sensitive to in-plane displacement aligned with the axis of the cavity the displacements caused by AE events are more easily interpreted than with standard AE. This is significant because the major damage mechanism of concern in unidirectional fiber reinforced composites is the breakage of the load bearing fibers. By aligning the axis of the cavity with the direction of the fibers the in-plane displacement and deformation experienced by the fibers should be readily detected.

The EFPI is also capable of "recording" strain events. While the initial gap distance of the fiber sensor is calibrated after the sensor is placed the fibers are free to float in the cavity. Any events that cause the gap distance between the fibers to increase or decrease will affect the steady state phase difference and be detected when the system is turned on. While the EFPI monitoring system can be easily adjusted to account for any changes in the gap distance, without reducing sensitivity, the amount of adjustment required is indicative of the cumulative level of deformation the material has experienced. This unique feature is very useful in monitoring the health of a structure because AE detection is a passive monitoring method and the system must be ready to observe and capture data when an event occurs since events due to damage are irreproducible.

The EFPI is impervious to electromagnetic interference (EMI) caused by passing automobiles or other common EMI around bridges. Furthermore, the EFPI can be applied to a wide variety of materials and surfaces, which is not true of many commercially available sensors. For instance, the NIST conical transducer requires a conductive path between the tip of the piezoelectric cone and the main signal amplifier in order for the sensor to be grounded and provide reliable information. While it is possible to prepare the surface of a non-conductive composite to accommodate the NIST transducer, it is impractical on many composite structures.

## 4 Methods and Materials

To test the accuracy and usefulness of the fiber sensor on structural composites several structures were instrumented with the EFPI. A pencil lead break (PLB) from a 0.3 mm HB lead in a Pentel Pencil with a lead break hood as specified by Higo [5] was used as the AE source for assessing the system response.

Structural shapes made of EXTREN 500 series fiber-reinforced materials were used in the experiments. EXTREN is a product manufactured by Strongwell that consists of unidirectional oriented fiberglass in an isophthalic polyester resin matrix. The wave velocity of longitudinal and shear stress waves through the material were determined by placing sender and receiver transducers on small plug samples with squared sides and polished sides. An aluminum, grade 6061, 330.20 mm by 330.20 mm square plate, 3.175 mm thick, was used to characterize both the sensor and the system response through comparison with displacement predicted by the DFEM [2].

A 152 mm by 152 mm by 920 mm EXTREN box beam with 15 mm thick walls was also instrumented with an EFPI AE sensor and EFPI strain sensor while being loaded in four point bending. The EFPI sensors were all aligned with the axis of the reinforcing fibers. A Physical Acoustics WD sensor was placed next to the EFPI AE sensor in order to compare the in-plane and out of plane results from testing. The data and resulting waveforms from each experiment were subsequently filtered using the MATLab Signal Processing Toolbox.

## 5 Comparison of DFEM model and EFPI Sensors

The response of the EFPI and the monitoring system were calibrated using a model. A dynamic finite element model was run with the test plate and sensor configuration in order to compare the specific components of displacement on an aluminum plate to the response of the EFPI. According to Prosser [2] the model gives a very close approximation to a point unload or pencil lead break (PLB) and the components of displacement from all three orthogonal directions can be directly calculated. However, the model does not exactly simulate the effect of a pencil lead break due to cell size limitations and the specific shape of the pencil lead at the time of a break.

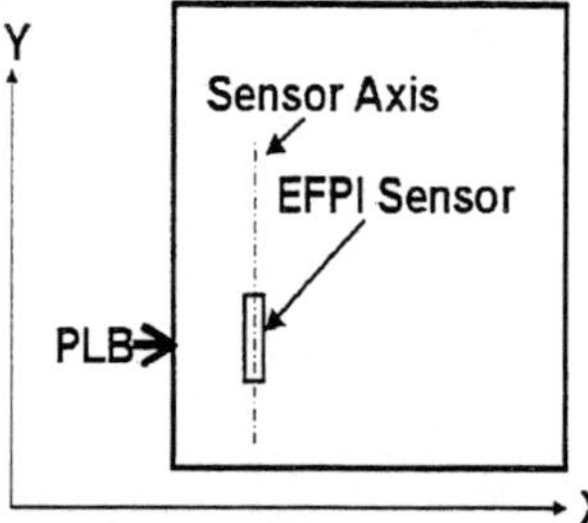


**Figure 1: Test Plate Configuration**

The PLB source for both the model and the test plate was positioned 63.5 mm in the Y direction from the bottom corner of the test plate. The center of the sensor in both the model and the test plate was located 63.5 mm in the Y direction and 63.5 mm in the X direction from the bottom corner of the test plate. Figure 1 shows the test configuration for the edge break. The data from the test plate was sampled at 8 MHz and the data from the model was generated at approximately 26.5 MHz. The results from the model and the EFPI sensor for a PLB located in the middle of the edge face are shown below (see Fig. 2 and 3). The data for both the model and the experiment was filtered using a Butterworth IIR bandpass filter. The pass band for each data set was 500 kHz to 1000 kHz.

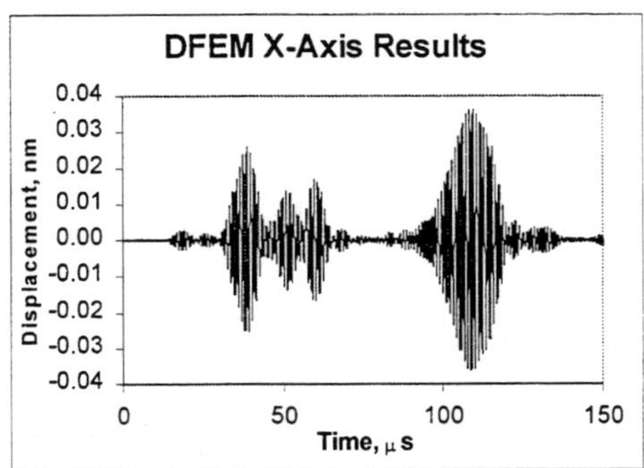

**Figure 2: Model results for in plane X-axis displacement**

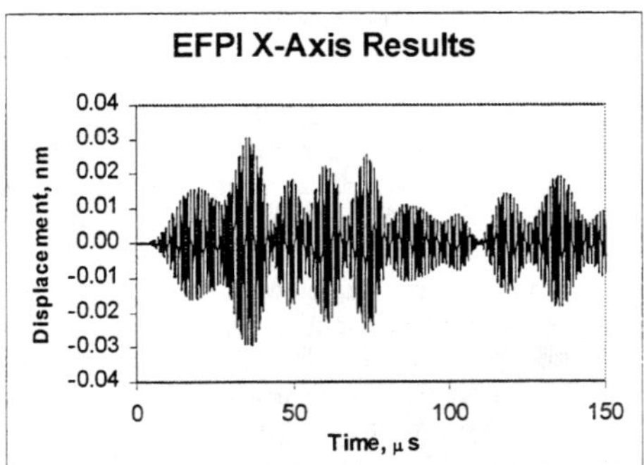

**Figure 3: Experimental EFPI sensor results for X-axis PLB**

The EFPI compares fairly well to the early part of the model response, however, after 70 us large differences appear. The difference between the model and the sensor could be caused by the difference in source size, or deviation of the pencil lead point break from the exact center of the edge face.

## 6 Results of EXTREN Plug testing

In order to characterize the material used in the experiments, the longitudinal and shear wave velocities were determined using small EXTREN samples. The wave speed for longitudinal waves in both along the fiber axis and transverse to it, shear and surface waves were determined experimentally using the appropriate transducers. The experimental results were compared with theoretical predictions using assuming an infinite, isotropic, elastic solid. The values for Young's Modulus, Poisson's Ratio and density used to calculate the wave velocities were, E = 17900 N/mm$^2$, $\mu = 0.2$ and $\rho = 1940$ kg/m$^3$ respectively. The wave velocity of each type of wave was calculated theoretically using the intrinsic material properties of the EXREN and the relationships that are described in the Handbook of Acoustics [3]. Equation 1 calculates the longitudinal wave velocity.

$$V_l = \{\frac{E(1-\mu)}{\rho(1+\mu)(1-2\mu)}\}^{1/2} \quad (1)$$

A wave velocity of 3181 m/s was observed along the fiber axis and 2787 m/s transverse to it. The values correspond to 99.3% and 87% of the theoretical value. Equation 2 calculates the shear wave velocity.

$$V_s = \{\frac{E}{2\rho(1-\mu)}\}^{1/2} \quad (2)$$

A wave velocity of 2399 m/s was observed which compares well to the theoretical value of 2401 m/s. Equation 3 calculates the Rayleigh wave speed based on an approximation using the shear wave velocity. A wave speed of 2189 m/s was obtained using the approximation.

$$V_r = V_t\{\frac{0.87\mu + 1.12\mu}{1+\mu}\} \quad (3)$$

It was assumed that internal friction would be the major source of signal loss in the material due to the nature of fiber reinforced composites. Observing the signal loss as a function of distance yields a good approximation of the internal friction characteristics in the composite material. According to Pollack [6] the attenuation coefficient in fiber-reinforced composite materials is 10 to 30 dB/m at 140 kHz. Piezoelectric transducers were used to measure the loss of signal amplitude as a function of distance on an EXTREN[tm] box beam. A 0.3 mm HB PLB was used as the source for sensor test. Physical Acoustics WD AE sensors monitored a series of pencil lead breaks. For each event, a different PLB source location and sensor location was chosen in order to determine the signal loss from a 3 dimensional distance away from the sensor. Several kinds of signal "flight paths" were chosen in order to determine the signal loss due to distance in a straight line from the sensor with an out-of-plane and an in-plane source. The signal loss for a source wave that traveled around the corners of the tube as well as one that traveled in plane but transversely to the fiber axis was obtained by using a series of sensor positions and lead breaks. A reference value of 1 mV was used to calculate the relative amplitudes.

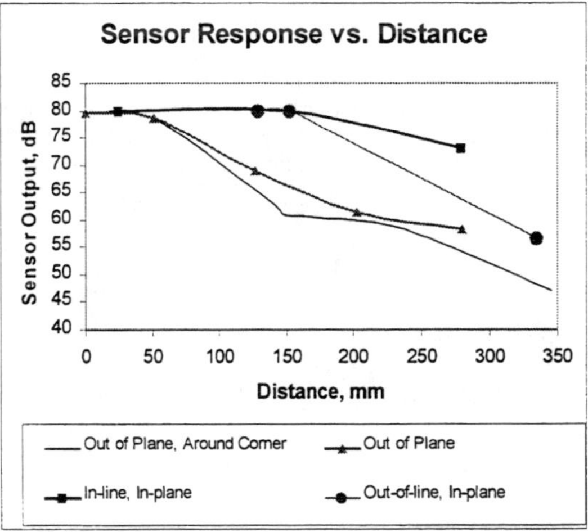

**Figure 4: Signal loss as a function of distance**

As expected, the least signal loss in the material occurred when the AE source was aligned with the reinforcing fiber axis. At a signal loss rate of 16.7 dB/m, sensors could be mounted as far as two meters apart before significant signal loss would make it difficult to detect AE events along the axis of the fibers. To accurately observe transverse signals however, sensors should be placed 100 – 200 mm apart across the fiber axis. Given the rate of signal loss from out of plane sources, it is preferable to monitor the in-plane disturbances because they travel farther and faster with less signal loss in this material.

## 7 Results of EXTREN Tube AE testing

A 152 mm by 152 mm by 920 mm EXTREN box beam with 15 mm thick walls was also instrumented with an EFPI AE sensor and EFPI strain sensor while being loaded in four point bending. The EFPI sensors were all aligned with the axis of the reinforcing fibers. A Physical Acoustics WD sensor was placed next to the EFPI AE sensor in order to compare the in-plane and out of plane results from testing. The sensors mounted on the tube during loading were synchronized and set at a 70 dB threshold. They were also set using a 50 μs pre-trigger.

Figures 5 and 6 are the filtered waveform of a single AE event that was simultaneously recorded by the PAC WD sensor and the EFPI. The signals were filtered using a Butterworth IIR bandpass filter, using a passband from 500 kHz to 1000 kHz, in order to eliminate most of the signal generated from matrix cracking. The sensors that observed the AE were 90 mm away from one of the loading contact points.

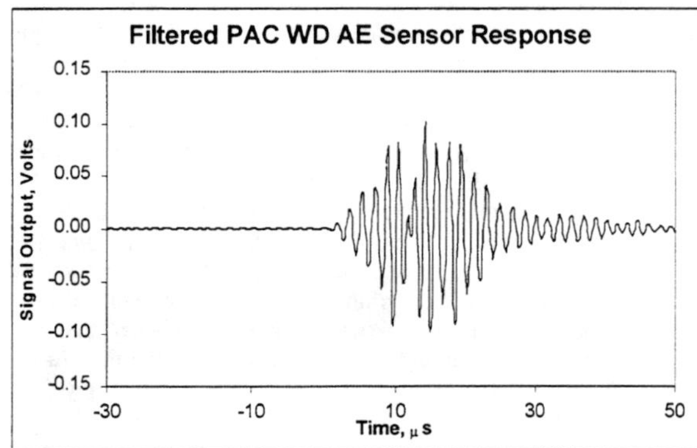

Figure 5: PAC WD sensor response to AE from beam loading experiment

The first peak of the PAC WD AE sensor in Fig. 5 directly corresponds to the time that the Rayleigh wave would have arrived at the sensor if it came from the center of the loading point. The first peak of the EFPI AE sensor in Fig. 6 directly corresponds to the time that the longitudinal wave aligned with the axis of the optical fiber would have arrived at the sensor if it came from the center of the loading point.

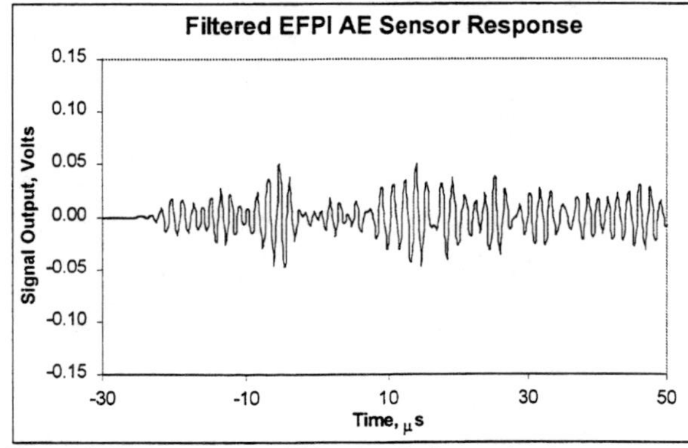

Figure 6: EFPI sensor response to AE from beam loading experiment

Given the complex nature of the material, it is not possible to clearly delineate whether or not the AE resulted from a fiber rupture, but given the speed and amplitude of the event it is highly likely. Figure 7 overlays both signals to directly compare the different components of the wave.

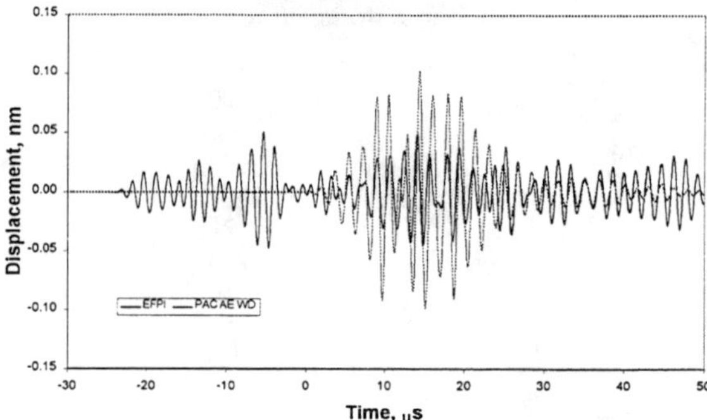

Figure 7: PAC WD signal overlaid on the EFPI output

## 8 Conclusions

This paper presents the results of a study to develop an efficient FRP composite infrastructure health monitoring system. The EFPI sensor is ideally suited for use in AE monitoring of composites in infrastructure applications. The EFPI sensor may give more detailed and easily interpretable information than standard piezoelectric sensors because of the nature of wave propagation through fiber reinforced material. Furthermore, in-plane displacements are easier to detect at farther distances in fiber reinforced materials than out of plane disturbances. The optical fiber sensor is able to detect AE events that occur along the major axis of reinforcement making it ideal for use on FRP. Future work will be done on AE signals from larger and more complicated FRP structures as well as FRP wrap applied to concrete columns and reinforced beams.


### Acknowledgements

The authors would like to thank W.H. Prosser for his modeling efforts and his help in interpreting the data from the model. The authors are also especially grateful to Mr. Jason Dietz and for his help with the monitoring equipment and the sensors. Finally, much of this effort would not have been possible without the help of M. Horne.


## 9 References


1 American Society of Civil Engineering Website, http://www.asce.org/govnpub/start.html, ASCE's 2001 Report Card on America's Infrastructure

2 Prosser, W.H., Hamstad, M.A., Gary J., O'Gallagher, A., 1999, "Reflections of AE Waves in Finite Plates: Finite Element Modeling and Experimental Measurements," Journal of Acoustic Emission, vol. 17(1-2), pp. 37-47.

3 Handbook of Acoustics, "Chapter 58: Structure-Borne Energy Flow", pp. 713 – 723, Ed. By Crocker, W., Pavic, G., John Wiley and Sons, New York.

4 Pollock, A.A., Handbook of Non-Destructive Engineering, Vol.6, "Acoustic Emission Inspection", pp. 278 – 294, John Wiley & Sons, New York.

5 Higo Y., Inaba H., "The General Problems of AE sensors," Acoustic Emission: Current Practice and Future Directions, ASTM STP 1077, W Sachse, J Roget, K Yamaguchi, Eds., American Society for Testing and Materials, Philadelphia 1991.

6 Pollock, A.A., "Classical Wave Theory in Practical Acoustic Emission Testing," Progress in Acoustic Emission III, The Japanese Society of NDI, Technical Report, 1986.

Proceedings of
2001 ASME International Mechanical Engineering Congress and Exposition
November 11–16, 2001, New York, NY
NDE-Vol. 21

IMECE2001/NDE-25812

# BLIND IMAGE RESTORATION FOR ULTRASONIC C-SCAN USING CONSTRAINED 4TH ORDER CUMULANTS

Uvais Qidwai
Chi-Hau Chen
ECE Department
University of
Massachusetts, Dartmouth
Email:
g_uqidwai@umassd.edu

## ABSTRACT

C-Scans are ways to visualize the sample under study by utilizing the reflections from various levels of non-homogenous wave transfer within the sample. Unlike other imaging techniques, C-scan are usually constructed from the pulse-echoed A-Scans by mapping the 1D windowed signal into a point corresponding to a pixel on the C-scan by calculating its energy within the window. Hence, although the A-scans are predominantly Gaussian in nature, the spatial 1D waveforms mapping into spatial energy values, results in completely unpredictable statistical characteristics. Also, the medium characteristics incorporate distortions that are essentially due to a non-minimum phase system response. Hence, the usual Second Order Statistics (SOS) based identification and deconvolution, i.e., correlation and covariance based techniques, may not work very well in this case.

In this paper, an approach is presented to use 4th order cumulants to deconvolve the effects of blurring in the C-Scans due to above-mentioned effects. The proposed approach is completely blind to the source or the type of distortion and the formulation is purely two-dimensional. When the blurring function is modeled as an Auto Regressive (AR) process, the image is restored recursively with the application of the inverse filter based on the AR estimate. A significant improvement in the image quality has been demonstrated. Especially, the edges are detected more prominently than present in the original image. Very little or no post processing is needed to obtain the final image.

## INTRODUCTION

HOS are extensions beyond the well-known second order statistics (SOS), i.e. correlation and covariance functions.

When signals are non-Gaussian, the 1st and 2nd moments alone are not sufficient to reproduce the pdf for the signal under study [1,2]. While this is true for Gaussian signals, the higher moment are well known to perform better for non-Gaussian cases as the required parameter for other distributions require statistical moments beyond mean and variance. There are several general motivations behind the use of HOS in signal processing. These include techniques to:

1. Suppress additive colored Gaussian noise of unknown power spectrum;
2. Identify non-minimum phase systems or reconstruct non-minimum phase signals;
3. Extract information due to deviation from Gaussianity; and
4. Detect and characterize non-linear properties in signals as well as identify non-linear systems [1].

**Statistical moments**

Higher order statistics are specific combinations of higher order statistical moments. For a real stationary zero mean random process $\{g(k)\}$, statistical moments are defined as [3]:

$$m_p^g(k_1,k_2,\cdots,k_{p-1}) = E\{g(k)g(k+k_1)\cdots g(k+k_{p-1})\} \tag{1}$$

where $E\{\ \}$ is the statistical expectation operator. Therefore, the mean is the first order moment $m_1^g = E\{g(k)\}$, and the covariance is the second moment $m_1^g(k) = E\{g(k)g(k+k_1)\}$.

**General definitions**

For a non-Gaussian, zero mean stationary random process, $g(k)$, with finite moments, the cumulants can be expressed in terms of moments as

Thus, if the random process $g(k)$ has zero mean, the

$$c_1^g = m_1^g = 0 \tag{2}$$

$$c_2^g(k_1) = m_2^g(k_1) \tag{3}$$

$$c_3^g(k_1,k_2) = m_3^g(k_1,k_2) \tag{4}$$

and

$$c_4^g(k_1,k_2,k_3) = m_4^g(k_1,k_2,k_3) - m_2^g(k_1)m_2^g(k_3-k_2) - m_2^g(k_2)m_2^g(k_3-k_1) - m_2^g(k_3)m_2^g(k_2-k_1) \tag{5}$$

**ESTIMATING HIGHER ORDER STATISTICS**

In practice, a finite number of data samples are available. Hence the actual formulation for the moments and cumulants Given data samples {g(k), k=0, 1, 2,…,$N$}, the mean of the data is first subtracted, then the higher order moments are computed using [2]

$$\hat{m}_p^g(k_1,k_2,\cdots,k_{p-1}) = \frac{1}{N}\sum_{j=1}^{N} g(j)g(j+k_1)\, g(j+k_2)\cdots g(j+k_{p-1}) \tag{6}$$

The higher order cumulants are estimated using

$$\hat{c}_1^g = \hat{m}_1^g = 0 \tag{7}$$

$$\hat{c}_2^g(k_1) = \hat{m}_2^g(k_1) \tag{8}$$

$$\hat{c}_3^g(k_1,k_2) = \hat{m}_3^g(k_1,k_2) \tag{9}$$

and

$$\hat{c}_4^g(k_1,k_2,k_3) = \hat{m}_4^g(k_1,k_2,k_3) - \hat{m}_2^g(k_1)\hat{m}_2^g(k_3-k_2) - \hat{m}_2^g(k_2)\hat{m}_2^g(k_3-k_1) - \hat{m}_2^g(k_3)\hat{m}_2^g(k_2-k_1) \tag{10}$$

Similar expressions of order higher than four can be obtained in similar fashion. As reported in the literature (see for example [1] and [4]), 3rd and 4th order statistics are sufficient tools for most applications.

**2D HOS**

The theory behind 1D HOS has been established very well during past two decades [1,2], however, the extension to 2D has recently been started [5].

In this work a constrained subset for cumulant calculation is used, based on the 4-Neighbor model for pixels in an image.

For a finite sample set of a random signal, the definitions of Moments $M_i$, and Cumulants $C_i$ are defined as follows for $i = 2,3,4$:

$$M_1(m,n) = \frac{1}{N}\sum_{k_1=1}^{N}\sum_{k_2=1}^{N} g(k_1,k_2) \tag{11}$$

$$M_2(m,n,\sigma_1,\sigma_2) = \frac{1}{N^2}\sum_{k_1=1}^{N}\sum_{k_2=1}^{N} g(k_1,k_2)g(\sigma_1-k_1,\sigma_2-k_2) \tag{12}$$

$$M_3(m,n,k_1,k_2,\tau_1,\tau_2) = \frac{1}{N^3}\sum_{k_1=1}^{N}\sum_{k_2=1}^{N} g(k_1,k_2)\, g(\sigma_1-k_1,\sigma_2-k_2)g(\tau_1-k_1,\tau_2-k_2) \tag{13}$$

$$M_4(m,n,\sigma_1,\sigma_2,\tau_1,\tau_2,\rho_1,\rho_2)=\frac{1}{N^4}\sum_{k_1=1}^{N}\sum_{k_2=1}^{N}g(k_1,k_2)$$
$$g(\sigma_1-k_1,\sigma_2-k_2)g(\tau_1-k_1,\tau_2-k_2)g(\rho_1-k_1,\rho_2-k_2) \tag{14}$$

The cumulants can be expressed in terms of moments as

$$c_2^g(m,n,\sigma_1,\sigma_2)=m_2^g(m,n,\sigma_1,\sigma_2)-[m_1^g(m,n)]^2 \tag{15}$$

$$c_3^g(m,n,\sigma_1,\sigma_2,\tau_1,\tau_2)=m_3^g(m,n,\sigma_1,\sigma_2,\tau_1,\tau_2)$$
$$-m_1^g(m,n)[m_2^g(m,n,\sigma_1,\sigma_2)-m_2^g(m,n,\tau_1,\tau_2) \tag{16}$$
$$+m_2^g(m,n,\sigma_1-\tau_1,\sigma_2-\tau_2)]+2[m_1^g(m,n)]^3$$

$$c_4^g(m,n,\sigma_1,\sigma_2,\tau_1,\tau_2,\rho_1,\rho_2)=$$
$$m_4^g(m,n,\sigma_1,\sigma_2,\tau_1,\tau_2,\rho_1,\rho_2)$$
$$-m_2^g(m,n,\sigma_1,\sigma_2)m_2^g(m,n,\rho_1-\tau_1,\rho_2-\tau_2)$$
$$-m_2^g(m,n,\tau_1,\tau_2)m_2^g(m,n,\rho_1-\sigma_1,\rho_2-\sigma_2)$$
$$-m_2^g(m,n,\rho_1,\rho_2)m_2^g(m,n,\tau_1-\sigma_1,\tau_2-\sigma_2) \tag{17}$$

## HOS-BASED IDENTIFICATION

In this section, 4[th] order cumulants-based algorithm and its usage with Auto Regressive (AR) representation of the C-Scan image is presented. In order to utilize the time samples' analogy with the positions of pixels in the C-scan image, the causal 4-Neighbor model for pixels in an image has been used.

### Image modeling

The term image refers to a 2D light intensity function [6], denoted as a function $f$ of two spatial variables $(m, n)$. As part of basic relationships between these values for an image, *Neighborhood* of an image is an important concept. A pixel $p$ at coordinates $(m,n)$ has 4-neighbors at coordinates $(m-1,n-1)$, $(m,n-1)$, $(m+1,n-1)$, $(m-1,n)$. This is shown in Fig. 1.

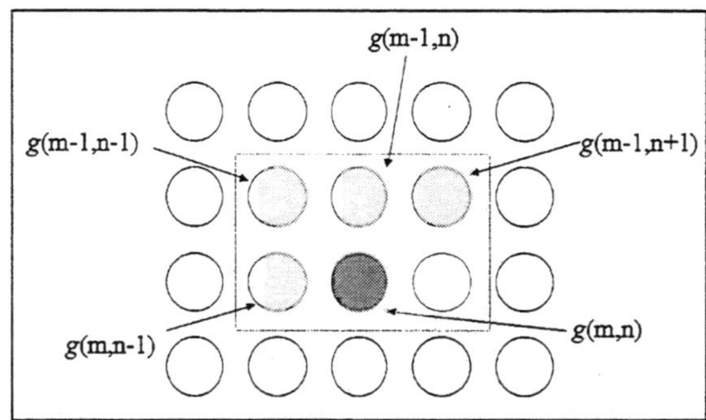


**Figure 1. Neighborhood of a pixel.**

The way an ultrasonic image is formed, it enforces a great deal of overlap between the neighboring particles, due to the vibrational behavior (spring model) of the molecular structure of the sample under study. A detailed study of such elastic wave behavior can be seen in the work of Yim and Sohn [7], confirming the idea that the neighboring pixels are interconnected in some manner that some characteristics are overlapped.

### 2D AR Model

In order to utilize the time samples' analogy with the positions of pixels in the c-scan image, the causal 4-Neighbor model for pixels in an image has been used, as shown in Fig. 1. The AR model for such arrangement can be written as follows:

$$g(m,n)=-[a_1g(m-1,n-1)+a_2g(m-1,n)$$
$$+a_3g(m-1,n+1)+a_4g(m,n-1)]+r(m,n) \tag{18}$$

where $g(m,n)$ is the measured pixel $g$ at position $(m,n)$ for the image under study, $a_i$ are the unknown parameters for AR coefficients, and $r(m,n)$ represents the superimposed noise pixel.

The system in (18) can be compactly represented as

$$g(m,n) = -[g(m-1,n-1) \quad g(m-1,n) \quad g(m-1,n+1) \quad g(m,n-1)]\begin{bmatrix} a_1 \\ a_2 \\ a_3 \\ a_4 \end{bmatrix} + v(m,n) \tag{19}$$

or

$$g = z\,\theta + v \tag{20}$$

Eq. (20) is called the normal equation in system identification literature and is a fundamental structure for any type of estimator design.

## 4<sup>TH</sup> ORDER-HOS-BASED ALGORITHM

Applying an approach similar to [8] on the image representation of Eq. (20), and assuming $g(m,n)$ to be a zero-mean, 4th order stationary, and non Gaussian random process, independent of $v(m,n)$

Pre-multiplying Eq. (20) by three shifted vectors of the measured samples, $z$, where $\delta_1 z$ represents the sample structure $z$ shifted by $\delta_1$ samples and $\delta_2 z$ represents the shift by $\delta_2$ and $\delta_3 z$ represents the sample structure $z$ shifted by $\delta_3$ samples. Hence Eq. (20) can be re-written as:

$$\delta_1 z \delta_2 z \delta_3 z \quad g^T = \delta_1 z \delta_2 z \delta_3 z \quad z^T + \delta_1 z \delta_2 z \delta_3 z \quad v^T \tag{21}$$

Taking expectations on both sides we get,

$$E\{\delta_1 z \delta_2 z \delta_3 z \quad g^T\} = E\{\delta_1 z \delta_2 z \delta_3 z \quad z^T + \delta_1 z \delta_2 z \delta_3 z \quad v^T\} \tag{22}$$

following definitions can be introduced,

$$\begin{aligned} \Phi &= E\{\delta_1 z \delta_2 z \delta_3 z \quad z^T\} \\ Y &= E\{\delta_1 z \delta_2 z \delta_3 z \quad g^T\} \\ R &= E\{\delta_1 z \delta_2 z \delta_3 z \quad v^T\} \end{aligned} \tag{23}$$

The right hand side of equations in (23) represents the 4th order moments. Hence Eq. (22) can be compactly written as:

$$Y = \Phi\theta + R \tag{24}$$

Under the conditions of assumptions, $\delta_1 z \delta_2 z \delta_3 z$ and $r$ are uncorrelated, and hence the solution for (24) can be given as follows:

$$\hat{\theta} = \Phi^{-1} \; Y \tag{25}$$

Once the model is obtained, the inverse filtering is performed after converting the low-pass (LP) structure of (19) into a standard high-pass form.

The estimate $\hat{\theta}$ can be obtained by using other estimators, like RLS (Recursive Least Square), LMS (Least Mean Square), MLE (Maximum Likelihood Estimator), etc…

As the 3rd and 4th order moments are calculated with almost the same procedure (except for the selection of samples etc…), a same strategy is presented in this section.

Since the model is constrained by the number of neighboring pixels, the estimated model is always 4th order, maximum. Unless, the non-causal 8-Neighbors are considered or special neighborhood is selected beyond 4-Neighbor model, the dimensions for the Φ matrix in the formulation would remain P×4, where P represents the number of pixels in the image.

A flow chart for this strategy is shown in Fig. 2.

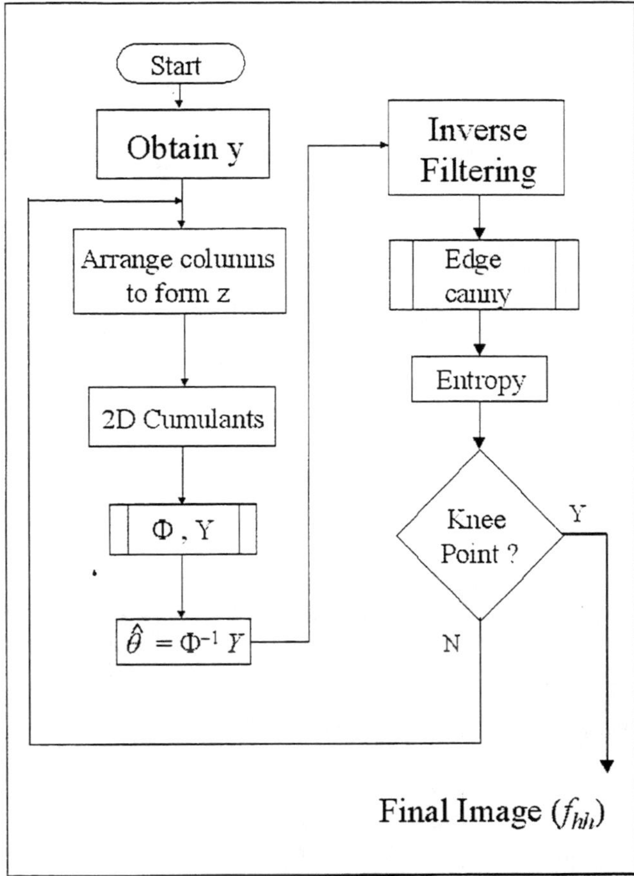


**Figure 2. Flow chart for the HOS-based scheme.**

The selection of $\delta_i$ would further constrain the calculation of moments. Within a neighborhood of 4, $i$ could be in the range of $i = 0, 1, 2, 3, 4$. Hence for a 4-Neighbor model, the samples are selected based upon the raster-scan (left-right, top-down) criterion. Hence utilizing the spatial coordinates as the modified index for the AR process model.

## RESULTS

The sample image used in this work is shown in Fig. 3. It represents three vertically drilled holes in an Aluminum blocks. Hole diameters are 12, 8 and 6 mm.

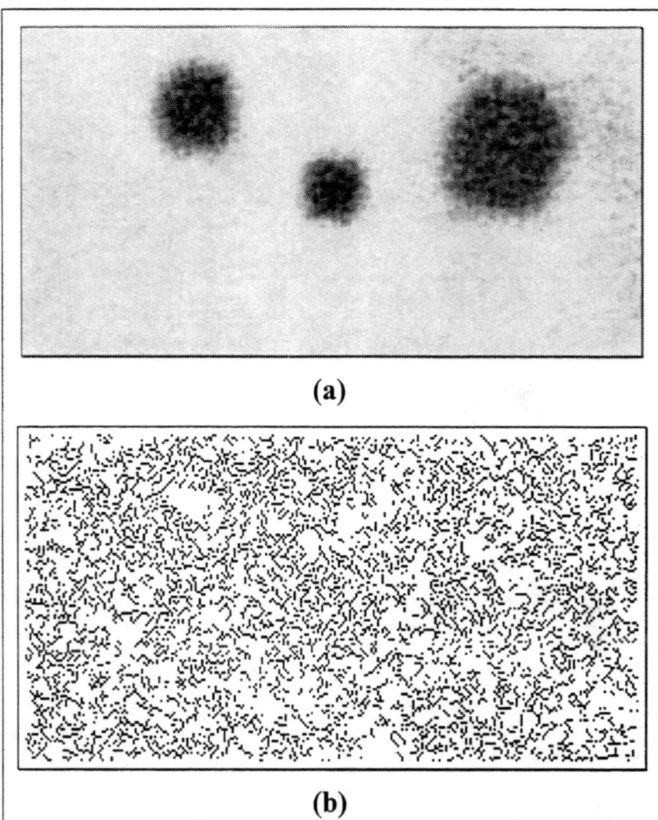

(a)

(b)

**Figure 3. (a) Original defect image, (b) Original edges**

The sample image in Fig. 3(a) is used with 4[th] order HOS. An iterative procedure for enhancement is utilized using Eq. (23) and (24) and using the Least Square estimate to solve Eq. (24), the image is progressively enhanced. The edges for first ten iterations is shown in Fig. 4.

### Loop Termination

One of the most interesting issues in image processing is to quantify the image in a way that certain conclusions related to the degree of improvement etc… be made. In this work, this is not only needed to compare the performance of the algorithms but also to use it as a decisive measure for loop termination. As can be seen in Fig. 2, before the decision for loop termination be made, two steps are carried out, i.e. Edge extraction and Entropy calculation. These are discussed in the following:

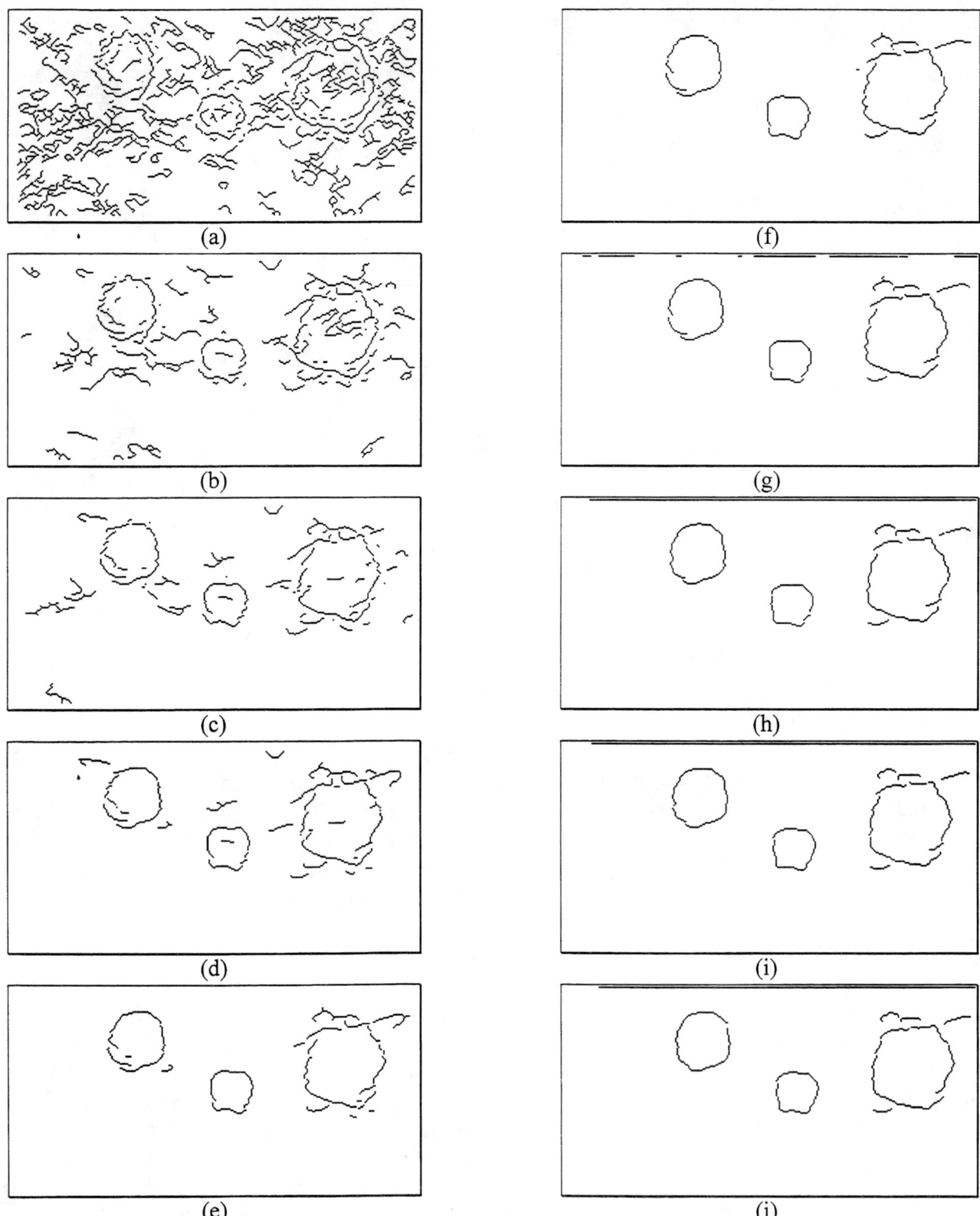


**Figure 4. Edges for each iteration, from 1 to 10.**

## Edge Detection

Edges are those places in an image that correspond to object boundaries. To find edges, places in the image where the intensity changes rapidly are sought, using one of these two criteria:

- Places where the first derivative of the intensity is larger in magnitude than some threshold
- Places where the second derivatives of the intensity has zero crossing edges providing a number of derivative estimators.

In this work, the Canny method [9] is used which finds edges by looking for local maxima of the gradient of the gray scale values. The gradient is calculated using the derivative of a Gaussian filter. The method uses two thresholds, to detect strong and weak edges, and includes the weak edges in the output only if they are connected to strong edges. This method is therefore less sensitive to noise, and more likely to detect true weak edges.

The Canny operator was designed to be an optimal edge detector (according to particular criteria - there are other detectors around that also claim to be optimal with respect to slightly different criteria). It takes as input a gray scale image, and produces as output an image showing the positions of tracked intensity discontinuities.

The Canny operator works in a multi-stage process. First of all the image is smoothed by Gaussian convolution which is a 2-D convolution operator that is used to 'blur' images and remove detail and noise. It uses a an isotropic (*i.e.* circularly symmetric) Gaussian kernel of the following form:

$$G(m,n) = \frac{1}{2\pi\sigma^2} e^{-\frac{m^2+n^2}{2\sigma^2}} \qquad (26)$$

where $\sigma$ represents the standard deviation. A typical kernel is shown in Fig. 5.

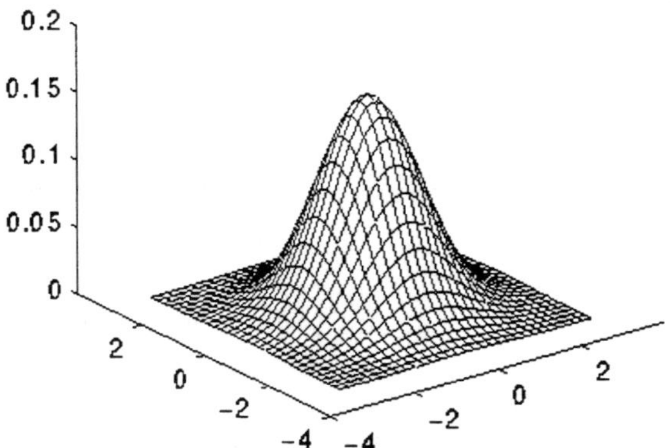


**Figure 5. 2D Gaussian Smoothing Kernel.**

Then a simple 2-D first derivative operator, as shown in Fig. 6, is applied to the smoothed image to highlight regions of the image with high first spatial derivatives.

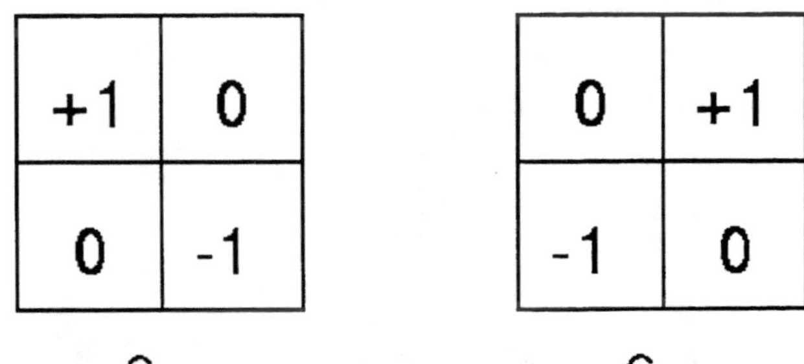

| +1 | 0 |
|---|---|
| 0 | -1 |

| 0 | +1 |
|---|---|
| -1 | 0 |

Gm Gn

**Figure 6. First order convolution masks.**

These kernels are designed to respond maximally to edges running at 45° to the pixel grid, one kernel for each of the two perpendicular orientations. The kernels can be applied separately to the input image, to produce separate measurements of the gradient component in each orientation (call these *Gm* and *Gn*). These can then be combined together to find the absolute magnitude of the gradient at each point and the orientation of that gradient. The gradient magnitude is given by:

$$|G| = \sqrt{G_m^{\ 2} + G_n^{\ 2}} \qquad (27)$$

The angle of orientation of the edge giving rise to the spatial gradient (relative to the pixel grid orientation) is given by:

$$\theta = \tan^{-1}\left(\frac{G_n}{G_m}\right) - \frac{3\pi}{4} \quad (28)$$

Edges give rise to ridges in the gradient magnitude image. The algorithm then tracks along the top of these ridges and sets to zero all pixels that are not actually on the ridge top so as to give a thin line in the output. The tracking process exhibits hysteresis controlled by two thresholds: $T_1$ and $T_2$, with $T_1 > T_2$. Tracking can only begin at a point on a ridge higher than $T_1$. Tracking then continues in both directions out from that point until the height of the ridge falls below $T_2$. This hysteresis helps to ensure that noisy edges are not broken up into multiple edge fragments.

**Entropy**

Entropy is defined as the degree of uncertainty in the image [6]. It is given by:

$$E = -\sum_{i=1}^{K} p_i \log p_i \quad (29)$$

Where $p_i$ represents the probability of the $i^{th}$ bin out of $K$ bins from the histogram mapping of the binary image.

The entropy is calculated in every iteration for the edges of the image that is obtained by using the Canny operator. A 2D histogram is first calculated which gives $K$ bins to be used by Eq. (29). Entropy is used in this work for dual purpose. Firstly, it gives a quantitative measure to represent the image improvement, and secondly it serves as a means to terminate the iterative procedure.

The entropy keeps decreasing as the image is improved. This corresponds to the reduction of uncertainties of blurring and noise. However, after certain stage, the entropy starts to increase due to the boundary effects in image reconstruction. This gives a knee-point that has been used as a terminal point to stop the iterations.

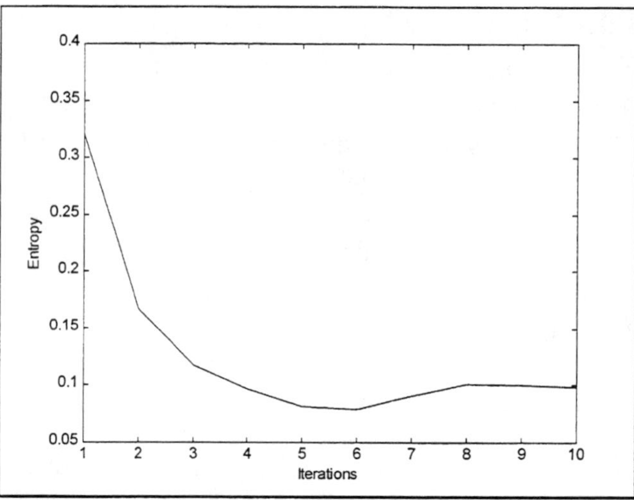


**Figure 7. Entropy curve for the edges in Figure 4.**

As can be seen from Fig. 7, the knee point occurs for the $6^{th}$ iteration. Hence the iterations are stopped at this point and the final image at this point is selected as the final image. This image is shown in Fig. 8.

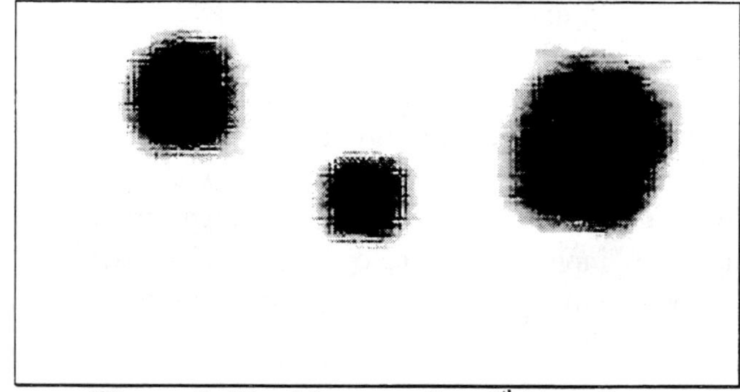

**Figure 8. Final image after $6^{th}$ iteration.**

## POST PROCESSING

Once the knee point is reached, the current image is declared as the final image. This image is already very improved version of the original distorted image. However, some post processing can be done to improve over certain issues like unwanted background removal, smoothing of torned edged, and contrast adjustment.

Firstly, the final edged image is operated upon by *dilation* operation, which is a morphological operation in image processing [6]. If any pixel in the input pixel's neighborhood is ON, the output pixel is ON. Otherwise, the output pixel is OFF. This helps in closing the boundaries and later on filling them up. This results in a Region of Interest (ROI) mask is obtained. When this mask is ANDed with the final image, it takes out the prospective regions in the image, thus removing any unwanted background. Hence from the final image, we obtain two complementary images, one representing the ROI and other the background. These two images are adjusted for the contrast in opposite directions and hence a better gray level distribution is obtained. The two images are then fused together to form the final image Fig. 9 shows the steps involved in this post-processing.

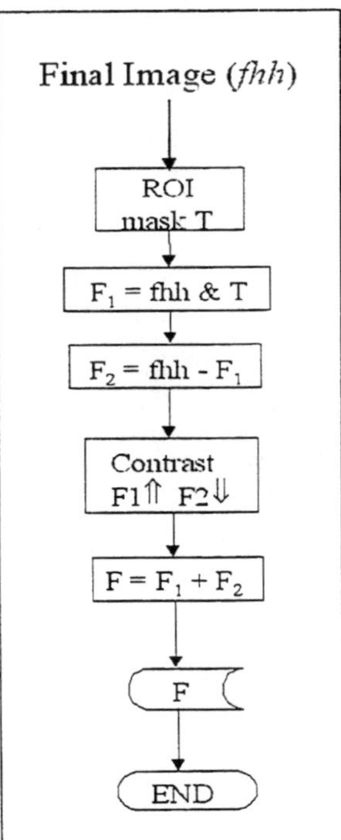


**Figure 9. Post-processing steps that can be utilized for fine tuning purposes.**

Using the post processing methodology of Fig. 9, the background is masked and the contrast adjustment is carried out. Hence the resulting image is shown in Fig. 10.

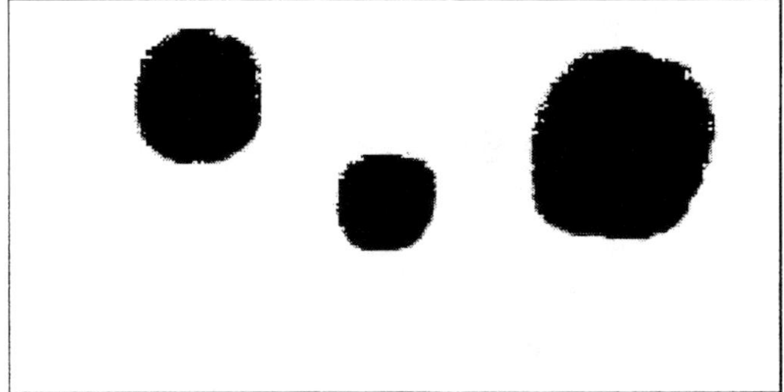

**Figure 10. Post processed image.**

## COCLUSION

In this paper, a new methodology has been presented to enhance the Ultrasonic C-Scans for NDE applications. The 4$^{th}$ order moments are calculated by multiplying the pixels of the image with the neighboring pixels and have been found to perform better than the second order statistics, like autocorrelation based techniques.

## REFERENCES


[1] Nikias, C. L., and Petropul, A. P., 1993, *Higher-Order-Spectra Analysis*, Prentice Hall Signal Processing Series, New York.

[2] Nandi, A. K., 1999, *Blind estimation using HOS*, Kluwer Academic Pblishers, New York.

[3] Nikias, C.L. and Raghuveer, M.R., 1987, "Bispectrum estimation: A digital signal processing framework", Proceeding IEEE, Vol. 75, No. 7, pp. 869-891.

[4] Mendel, J. M., 1991, "Tutorial on HOS in signal processing and system theory: theoretical results and some applications", Proceedings of the IEEE, Vol. 79, No. 3, pp. 278-305.

[5] Ibrahim, H. M., and Gharieb, R. R., 1999, "Estimating two-dimensional frequencies by a cumulant based FBLP method", IEEE

Transactions on Signal Processing, Vol. 47, No. 1, pp. 262-266.

[6] Gonzales, R. C., and Woods, R. E., 1992, *Digital Image Processing*, Addison-Wesley Publishing Company, New York.

[7] Yim, H., and Sohn, Y., 2000, "Numerical simulation and visualization of elastic waves using mass-spring lattice model", IEEE Transactions on Ultrasonics, Ferroelectrics, and Frequency Control, Vol. 47, No. 3, pp. 549-558.

[8] Inouye, Y., and Tsuchiya, H., 1991, "Identification of linear systems using input-output cumulants", Int. Journal of Control, Vol. 53, No. 6, pp. 1431-1448.

Proceedings of
2001 ASME International Mechanical Engineering Congress and Exposition
November 11–16, 2001, New York, NY
NDE-Vol. 21

**IMECE2001/NDE-25813**

# STOCHASTIC PROPERTIES OF ANISOTROPIC MATERIALS

Michael "Mick" Peterson
Assistant Professor of Mechanical Engineering,
University of Maine, Orono, ME, 04469-5711

Miao Sun
Graduate Assistant of Mechanical Engineering,
University of Maine, Orono, ME, 04469-5711

## ABSTRACT

Extensive research has been directed toward the development of methods for the optimal recovery of elastic properties from ultrasonic measurements. For a number of applications both the elastic and damping characteristics of the materials are required in design. The use of the optimal recovery does present challenges when applied to either man-made or natural anisotropic materials. In many cases manufacturing variability results in a need for a statistical description of the elastic and damping properties. In addition, errors in material lay-up or growth patterns may result in mis-orientation of the principle materials axes with respect to the geometrical axes. In this work, an examples is shown that demonstrates the recovery of the elastic properties of a natural material when stochastic properties are required. Statistical descriptions of the materials properties are obtained for wood of two different types of material. Results are shown assuming a nominally orthotropic orientation, although the existence of curvilinear coordinates is acknowledged..

## INTRODUCTION

The objective of this work is to recover the full elasticity tensor for juvenile and mature wood in ponderosa pine. The recovery of the elastic properties for the juvenile wood is important because ponderosa pine has potential commercial value in a full round configuration [1]. The importance of ponderosa pine is a result of the need for thinning of the National Forests after many decades of fire suppression. Most of this material is of a small diameter, less than 8 inches in diameter, and thus has a larger percentage of juvenile wood. The availability of elastic properties for this material can assist directing the material to its highest and best use which will enhance the commercial viability of the mechanical thinning that may be required in many areas as an alternative to control burns.

Previous work has been limited in the area of wood material characterization with respect to developing an understanding of the full elastic properties of wood. In recognition of the complexity of the wood as a material, bending modulus is often reported for dimension lumber. This is, of course, a property that is of great interest in many applications for wood and includes a significant effect of the shear stiffness in most configurations for wood beams. A more complete description of the material properties of small diameter timber is, however, useful since the loading of material in a number of potential applications may be more complex depending on the connection details and other application related details of the use of the material.

For these reasons as well as to support an improved understanding of the properties of wood as a material, this work is focussed on the recovery of both shear and normal properties of the material in all three material axes of the material.

## THEORETICAL BACKGROUND

### Elastic Symmetry of Wood

As in much of the other literature of wood-based composites, in this work the material will be assumed to be orthotropic [2]. In orthotropic symmetry the material posses three mutually perpendicular planes of symmetry. In contrast, the overall tree is transversely isotropic. For transverse isotropy, the material has an axis of symmetry, the central axis of the tree, such that all planes that contain this axis have equivalent material properties.

The assumption of orthotropic symmetry is a particularly good assumption for lumber that is derived from old growth timber. As the radius from the center of the tree increases, symmetry with respect to the plane that has the radius as the unit normal becomes increasingly reasonable. However, for small diameter timber, the plane that contains the growth rings has a radius of curvature that is less than four inches. In the case of small diameter timber this is not actually a plane of symmetry in a Cartesian coordinate system (Figure 1). If a curvilinear coordinate system is used, the plane of symmetry continues to exist even in small diameter timber. This effect will be neglected in the present work, although it should be recognized that the symmetry assumed in the experiments is only strictly applicable after a general tensor transformation has been performed into an appropriate coordinate frame [3]. Well-established techniques for addressing the complexity of non-Cartesian coordinates exist using generalized tensors. The use of generalized coordinates does, however, introduce significant complexity into the processing and analysis of the data presented.

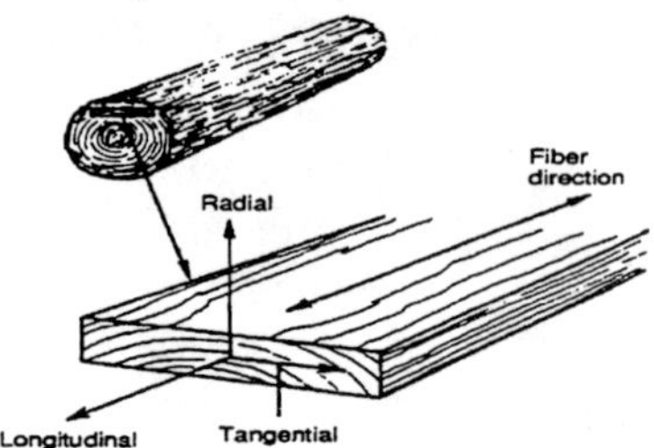


**Figure1. Typical dimension lumber obtained from round wood.**

The samples used for the measurements are cut based on visual identification of material axes. All of the ultrasonic measurements are assumed to occur along axes of symmetry. Some insight, albeit an incomplete description, into the applicability of these assumptions will be provided through the natural redundancy of the measurements that are made in a simple orthotropic material. The error between the measurements of the same constants made in orthogonal directions allows the assumptions to be tested to a certain extent. While the similarity of the repeated measurements is surely indicative of the existence of orthotropic material properties, it is certainly not a sufficient condition for this minimal level of symmetry to exist.

In the measurements, a Cartesian coordinate system is used, with three normal directions defined in the conventional wood grain directions. The direction along the wood grain (the fiber direction) is called the longitudinal or designated as the $X_1$ direction. The radial direction in the same orientation as the wood rays is defined as the radial or $X_3$ direction. The tangential or the $X_2$ direction is so designated since it is tangent to the annual growth rings and represents the third basis vector (Figure 1). Then the growth rings in Figure 1 are modeled as planar growth layer. The result is clear that wood which is modeled as an orthotropic material with material axes $(X_1, X_2, X_3)$ are coincident with geometric axes $(L,T,R)$ (Figure 2).

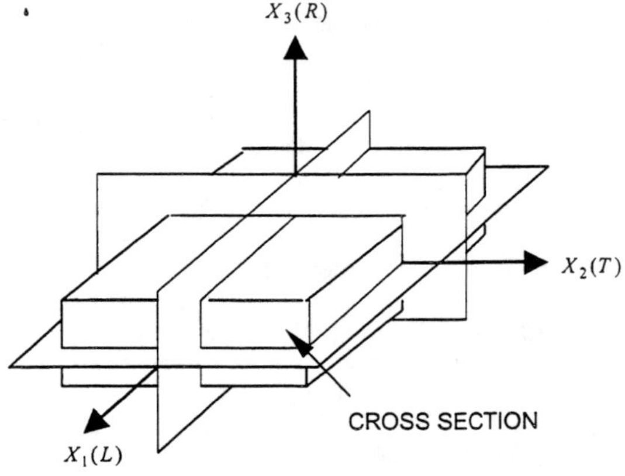


Figure2. Orthotropic axes in a block of clear wood

## Ultrasonic determination of elasticity tensor from wavespeed measurement

The ultrasonic determination of material elasticity constants is well established. Works by Van Buskirk [4], Norris [5], Aristegui & Baste [6] and Bucur [7] are combined in order to apply the methods to characterization of juvenile and mature wood.

For an orthotropic material, the elastic tensor contains nine independent elastic components that must be determined. From the equations of motion in a continuum:

$$\sigma_{ij,j} = \rho \ddot{u}_i \qquad (1)$$

where standard summation convention is assumed on the subscripts, $i,j,k,l$ with a comma denoting differentiation with respect to the spatial index following the comma. A superior dot is then used to indicate the differentiation with respect to time. In addition the terms are: $\sigma_{ij}$ is stress tensor, $\rho$ is the density of the material, $u_i$ is the first order tensor for the displacements

For corresponding linearly elastic solid the constitutive equation is

$$\sigma_{ij} = C_{ijkl}\varepsilon_{kl} \qquad (2)$$

where $\varepsilon_{kl}$ is strain tensor, $C_{ijkl}$ is the fourth order elasticity tensor for the material.

For an assumption of small displacement, the displacement equation of motion is:

$$C_{ijkl}u_{k,jl} = \rho \ddot{u}_i \qquad (3)$$

where $u_{k,jl}$ is the second partial derivative of displacement with respect to coordinate index $j$, and $l$. $\rho$ is the density of the material.

Assuming a solution of the form

$$u_i = A_0 \alpha_i e^{i[k(l_i x_i - \omega t)]} \qquad (4)$$

where $A_0$ is the incident amplitude which will be divided into components that propagate in directions that are determined by the relationship between the incident angle and the material symmetry axes.

Substituting *(4)* into *(3)* gives an eigenvalue equation (Christoffel s equation) which is a cubic polynomial in $V^2$.

$$(C_{ijkl}n_l n_j - \delta_{ik}V^2\rho)P_m = 0 \qquad (5)$$

where $P_m$ is the polarization vector, $V$ is the phase velocity of ultrasonic wave, $\delta_{ij}$ is the Kronecker delta symbol, $n$ is the wave propagation direction with components of $n_1, n_2, n_3$.

In order to have a nontrivial solution of the homogeneous system of equations, the determinant of the coefficient matrix must vanish:

$$Det[\Gamma_{ij} - \delta_{ik}V^2\rho] = 0 \qquad (6)$$

where $\Gamma_{ij} = C_{ijkl}n_l n_j$ is called Christoffel s tensor. For an orthotropic material the elements are given by:

$$\begin{aligned}
\Gamma_{11} &= n_1^2 C_{11} + n_2^2 C_{66} + n_3^2 C_{55} \\
\Gamma_{22} &= n_1^2 C_{66} + n_2^2 C_{22} + n_3^2 C_{44} \\
\Gamma_{33} &= n_1^2 C_{55} + n_2^2 C_{44} + n_3^2 C_{33} \\
\Gamma_{23} &= n_2 n_3 (C_{23} + C_{44}) \\
\Gamma_{13} &= n_1 n_3 (C_{13} + C_{55}) \\
\Gamma_{12} &= n_1 n_2 (C_{12} + C_{66})
\end{aligned} \qquad (7)$$

where for convenience, $C_{ijkl}$ are simplified to $C_{mn}$ according to the following notation :

$$\begin{aligned}
&23, or, 32 - 4 \\
&13, or, 31 - 5 \\
&12, or, 21 - 6
\end{aligned}$$

Therefore, for orthotropic and transversely isotropic materials, in any given direction, three possible acoustic waves can be generated: a quasi-longitudinal wave and two polarization of quasi-transverse waves. The phase velocities of all of these waves are known functions of the elastic properties and the density of the material.

As a result of material symmetry waves propagate along the longitudinal, tangential or radial directions of the three material symmetry axes, in all three directions, all three wave types can be generated. The directions chosen are the material axes, which are also called the "specified directions" by some authors who have described the problem in a more general sense [8].

A Cartesian coordinate system with axes $X_1, X_2, X_3$ parallel to the edges of the cube is assumed. For wave motion in $X_1$, the longitudinal direction in the wood where $n_1 = 1, n_2 = n_3 = 0$, one longitudinal wave and two transverse waves were generated. The two transverse waves propagate along the specific direction and have displacement polarizations that are orthogonal. These two polarizations correspond to separate wave velocities that are associated with different terms of the elasticity tensor. (Figure 3 and Figure 4)

$$\begin{aligned}
V_{11}^2 \rho &= C_{11} \qquad (8-1) \\
V_{12}^2 \rho &= C_{66} \qquad (8-2) \\
V_{13}^2 \rho &= C_{55} \qquad (8-3)
\end{aligned}$$

where $(8-1)$ represents longitudinal wave, $(8-2)$ and $(8-3)$ represent transverse wave with $X_2$ and $X_3$ polarization respectly.

Table 1. shows the subscript notation used for the $V$.

By obtaining wave velocity data from $X_2$ and $X_3$ direction, all the six diagonal terms of elasticity tensor can be obtained directly. The terms of the tensor that have been obtained up to this point are all diagonal terms of this type.

| | |
|---|---|
| $V_{ii}$ | velocity of a longitudinal wave traveling in the $x_i$ direction |
| $V_{ij}$ | velocity of a transverse wave traveling in the $x_i$ direction with particle motion in the $x_j$ direction |
| $V_{ij/ij}$ | velocity of a longitudinal wave with motion in $i-j$ plane |

Table1. Notation used to describe the direction and mode of vibration of the wave velocity

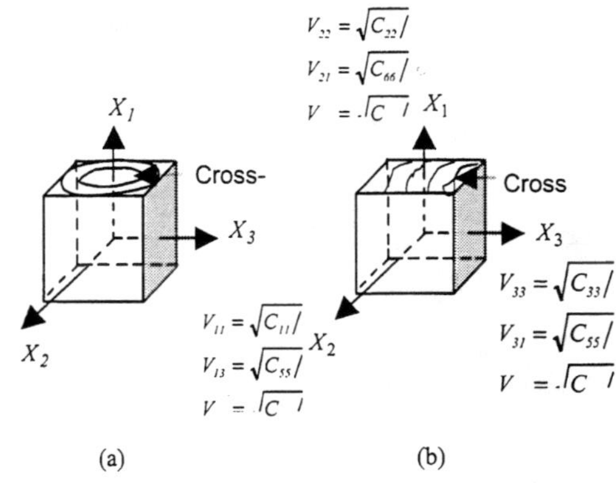


Figure3. Ultrasonic wave velocity in an orthotropic wood, (a). propagation in juvenile wood. (b). propagation in mature wood

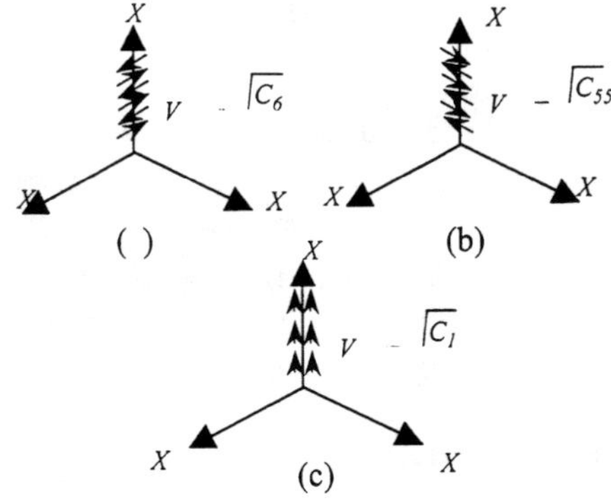


Figure 4. Wave propagation in $X_1$ direction of an orthotropic solid, (a) $X_2$-polarized, (b) $X_3$-polarized, (c) $X_1$-propagation

The three off-diagonal terms require that a wave be propagated in a direction other than a principal material axis. Instead of propagating along the $X_1, X_2, X_3$ direction, let wave propagating at $45^0$ angle to the axis in the $X_1 - X_2$, $X_2 - X_3$, and $X_1 - X_3$ plane, the direction vectors for the waves then become $(\frac{\sqrt{2}}{2},\frac{\sqrt{2}}{2},0)$, $(0,\frac{\sqrt{2}}{2},\frac{\sqrt{2}}{2})$ and $(\frac{\sqrt{2}}{2},0,\frac{\sqrt{2}}{2})$. The wave velocities obtained in these directions are designated by $V_{ij/ij}$, would represents longitudinal wave propagation. Transverse waves would be designated by $V_{i/j}$, representing their respective polarization vectors. All the three off-diagonal terms can be determined by (9).

$$C_{12} = \frac{\sqrt{\left(C_{11}n_1^2 + C_{66}n_2^2 - V_{12/12}^2\rho\right)\left(C_{66}n_1^2 + C_{22}n_2^2 - V_{12/12}^2\rho\right)}}{n_1n_2} - C_{66}$$

$$C_{13} = \frac{\sqrt{\left(C_{11}n_1^2 + C_{55}n_2^2 - V_{13/13}^2\rho\right)\left(C_{55}n_1^2 + C_{33}n_2^2 - V_{13/13}^2\rho\right)}}{n_1n_3} - C_{55} \qquad (9)$$

$$C_{23} = \frac{\sqrt{\left(C_{22}n_2^2 + C_{44}n_3^2 - V_{23/23}^2\rho\right)\left(C_{44}n_2^2 + C_{33}n_3^2 - V_{23/23}^2\rho\right)}}{n_2n_3} - C_{44}$$

## Velocity measurement method

The longitudinal waves required as long as the shear waves are all obtained using direct transmission technique (when the specimen is in contact with the longitudinal transducer). The specimens used are the same for both the longitudinal and transverse measurements. Fifteen specimens of juvenile ponderosa pine as well as twenty specimens of mature ponderosa pine are used. The longitudinal and shear modes are generated using $X-cut$ and $Z-cut$ piezoelectric transducers.

The longitudinal wave measurement is straightforward since only a single quasi-longitudinal wave propagates. However, due to the arbitrary particle displacement $u(t)$ of the transverse components (Figure 5), it is necessary to rotate the transverse wave transducer through $180^0$ to find the largest amplitude of transverse component that should correspond to $X_3$-polarization. From the maximum amplitude, the transducer is rotated $90^0$ to get another the orthogonal component that corresponds to the $X_2$-polarization. These two polarization correspond to the $C_{55}$ and $C_{66}$ components.

Once the particle oscillations and wave velocities for the two models of transverse wave propagation in the $X_1$ direction have been determined, the same testing procedure is repeated in the $X_2$ direction and $C_{44}$, $C_{66}$ components are obtained. The $C_{44}$ and $C_{55}$ components are found by propagating a wave in the $X_3$ direction.

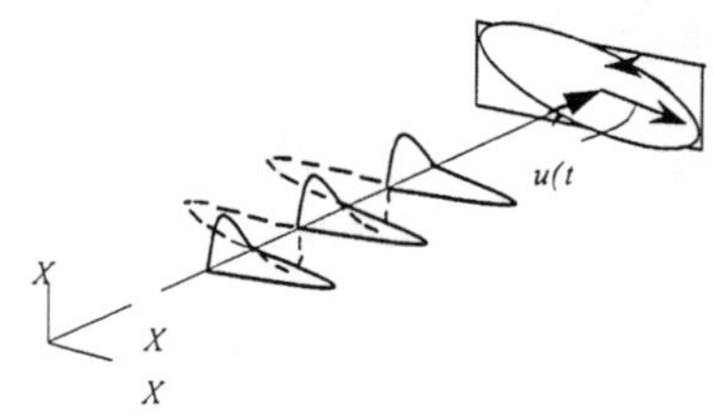


Figure 5. Combination of polarized transverse wave to form arbitrary particle displacement $u(t)$

## SPECIMEN PREPARATION

Using logs obtained from a prior testing section, cubes were harvested from the central portion of the tree. The middle section of the log was removed from the log using a horizontal band saw. $11/2''$ thick disks were cut from each end of the middle section, just beyond the break. From these disks, $1''$ cubes were harvested from the juvenile and mature portions of the log. It is assumed that the first 20 rings from the center of the disk represent juvenile wood. Figure 6 shows a picture of a cross section of a log. The cubes were then cut at appropriate angles from the axis of the disk. Figure 7 shows typical samples of cube locations.

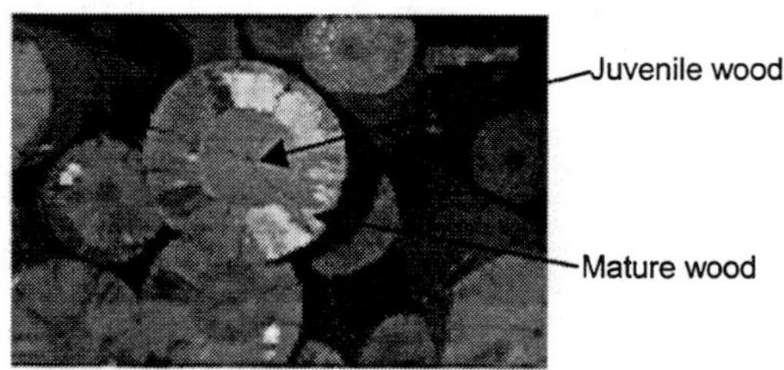


Figure 6. Cross-section of log showing portions of juvenile and mature wood

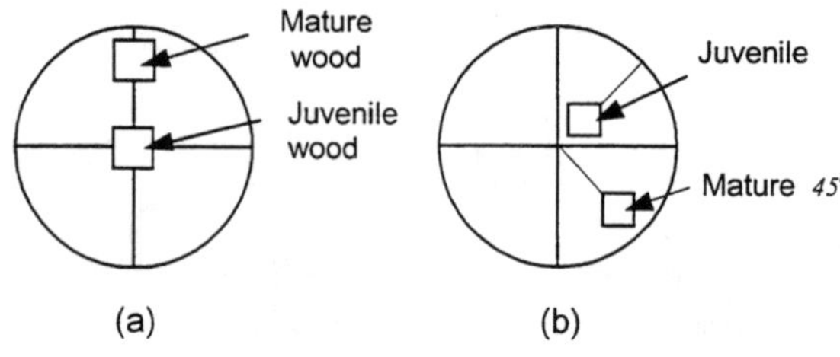


Figure 7. Cube orientation in T-R orthotropic plane, (a). No-rotation in TR plane,(b) Rotation in T-R plane

In order to obtain the three off-diagonal terms, additional $3-5''$ cylinders were cut. These sections were used to harvest cubes that are in a different orientation to the longitudinal axis. The cylinders were divided into four parts, cutting a cross through the center of the log section in the longitudinal axis. Cubes were then harvested from each quadrant. Figure 8. shows the orientation of the cubes.

After the cubes were removed from the log, the specimens were polished with a belt sander. The belt sander was used to obtain flat and smooth faces and to ensure parallel sides. The dimensions of the cubes were then taken and recorded. The cubes were conditioned prior to testing normalize the moisture content at approximately 15% of fiber saturation.

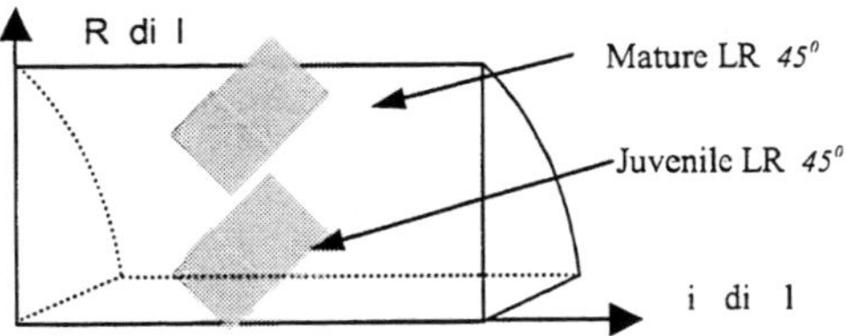


Figure 8. Rotation of cube specimen in L-R orthotropic plane

## DATA AND SIGNAL PROCESSING

The cross-correlation method was used to obtain accurate relative time delays.[9] The cross-correlation is used to estimate the time difference $(\tau)$ between two signals, a reference signal for a known material (aluminum) and an unknown sample (ponderosa pine).

Windowing of the signal is also required to truncate the additional wave paths that occur in the small block samples.

### Cross-correlation

The cross-correlation approach is used to determine the relative time delay $(\tau)$ between two waveforms. A reference signal which was acquired from an aluminum block $x(t)$ and wood block signal $y(t)$ are concerned, the cross-correlation between the two signals is a sequence $r_{xy}(t)$, which is defined as

$$r_{xy}(\tau) = \sum_{t=-N}^{+N} x(t)y(t-\tau) \qquad (10)$$

where $\tau$ is time shift or time delay parameter and the subscript $xy$ indicates the signals being correlated.

The cross-correlation method provides a measure of similarity between the power in the two signals. The relative time delay between the two signals is the time at which the cross-correlation $r_{xy}$ is a maximum. This time delay $\tau_{max}$ represents the relative shift.

The phase velocity ($V_{ij}$) of the ultrasonic waves propagating through the wood can be estimated by:

$$V_{ij} = (d_{wood} \times V_{al})/(d_{al} + (\tau_{ij\,max} \times V_{al}) \qquad (11)$$

where $V_{al} = 3200(m/s)$, which is the velocity of ultrasonic wave propagating in the reference specimen (aluminum), $d_{wood}$ and $d_{al}$ are the distances of wave propagating through cube and aluminum. $\rho$ is the density of wood.

The subscripts $i$ and $j$ follow Table 1 with $\tau$ representing the time delay.

Due to the size of the specimen used, the ultrasonic wavelength is on the order of the dimensions of the cube. As a result, in addition to the material dispersion that may be evident due to the visco-elasticity of the material, significant geometric dispersion should result as well. However, the use of the cross-correlation to determine the velocity of the ultrasonic wave results only in a measure of the relative difference in the group velocity between the reference sample (the aluminum block) and the unknown sample (the Ponderosa Pine) [9]. For the range of velocities considered, the only geometrical dispersion that would impact the measurement is the differences in the phase velocities between the reference sample and the unknown sample. As a result, unlike absolute measurements of velocity, the effect of dispersion even for this size sample can be assumed to be minimal and may be neglected.

### Signal windowing

To ensure that multiple reflections are not included in the time signal, a window is used to select the first arrival. The window is located at a point where the wave has traveled two complete passes through the specimen. Since the time delay is not known a-priori, it is necessary to use an adaptive calculation of the window position. To make an initial estimate of the time of flight, a phase velocity is initially estimated based on the reference velocity.

For the calculation used the initial estimates are shown in Table 2. The values were corrected to adjust the time bases on measurement of the two sets of specimens.

A simple truncation of the signal using a rectangular window introduces energy in the high frequencies. The increased energy in the high frequencies is a result of the abrupt transition to zero energy. A triangular window is used as a compromise between simplicity and minimization of the high-energy components. Specifically, if $w(t)$ is defined as a window function, then

$$w(t) = \begin{cases} 1 & (t < 0.8p) \\ (p-t)/(1-0.8) & (0.8 \le t \le p) \\ 0 & (t > p) \end{cases} \qquad (12)$$

where $p$ is the time of flight when the wave has made two complete passes through the specimen.

| | $V'_{11}$ | $V'_{22}$ | $V'_{33}$ | $V_{44}'$ | $V_{55}'$ | $V_{66}'$ | $V'_{12}$ | $V'_{13}$ | $V'_{23}$ |
|---|---|---|---|---|---|---|---|---|---|
| juvenile | 3800 | 1700 | 2300 | 600 | 1200 | 1200 | 1800 | 1800 | 1500 |
| mature | 5000 | 1500 | 2300 | 600 | 1200 | 1200 | 2000 | 2000 | 1600 |

Table 2. Initial estimated velocity for juvenile and mature wood

## RESULT AND VERIFICATION OF THE TECHNIQUE

Initial works have been done with the data and signal processing shown to determine the elasticity tensor. The spatially averaged values of the elasticity tensor measured from fifteen specimen of juvenile and twenty specimen of mature ponderosa pine are listed in Table3. It should be mentioned again that the coordinate system is the one that the longitudinal direction is labeled as $X_1$, the tangential direction is labeled as $X_2$ along with radial direction labeled as $X_3$. The comparison of elastic properties between juvenile and mature wood is also presented.

The transducer used to generate transverse wave was very sensitive to the relationship of the polarization of the wave to material axes. Verification of the shear wave technique is used to verify the measurements of the elasticity tensor. From the testing procedure , data for $C_{44}, C_{55}, C_{66}$ is obtained twice. This data should be nearly identical for an orthotropic material, the difference only represents the error in the measurement and deviation of the material from strict orthotropic symmetry. Thus, the comparison of these two sets of data can be used to estimate measurement uncertainty.

The comparison results are shown in Table.3.

For the juvenile wood, the values of the elasticity tensor $C_{55}$ measured from either the radial direction $(R)$ or tangential direction $(T)$ correspond well with those obtained from longitudinal direction $(L)$. For example, the average value of $C_{44}$ measured from $T$ direction is $0.5282 \pm 0.2357 (GPa)$, the corresponding value measured from $R$ direction is $0.5244 \pm 0.2.23\ (GPa)$ (Figure 9(a))

However, for the mature wood, case is not the same. From Table.3, it is seen the value of $C_{44}$ measured from $T$ direction, which is $0.3404 \pm 0.2530 (GPa)$, is significant different with that measured from $R$ direction, which is $0.5438 \pm 0.3871 (GPa)$. Indeed, the error is introduced by modeling tangential direction $(T)$ of the wood block as an orthotropic material. (Figure 9(b)). It is clear in the figure that due to the curvature of the rings the symmetry plane assumption in the tangential direction is a poor assumption.

| Average elasticity tensor | | Juvenile wood $(GPa)$ | Mature wood $(GPa)$ |
|---|---|---|---|
| $C_{11}$ | | 4.9095 (0.8425) | 10.627 (2.9269) |
| $C_{22}$ | | 2.2600 (0.7041) | 0.6819 (0.2534) |
| $C_{33}$ | | 2.2999 (0.6855) | 1.9731 (0.5744) |
| $C_{44}$ | measured in T direction | 0.5282 (0.2357) | 0.3404 (0.2530) |
| | measured in R direction | 0.5244 (0.2023) | 0.5438 (0.3871) |
| $C_{55}$ | measured in L direction | 1.4792 (0.8350) | 0.9927 (0.1650) |
| | measured in R direction | 1.4252 (0.7242) | 1.0140 (0.3575) |
| $C_{66}$ | measured in L direction | 1.4793 (0.8351) | 1.0418 (0.4209) |
| | measured in T direction | 1.5071 (0.7991) | 0.9957 (0.3721) |
| $C_{12}$ | | 1.5834 (0.6063) | 1.5462 (0.5646) |
| $C_{13}$ | | 2.0567 (0.8657) | 2.5312 (0.6016) |
| $C_{23}$ | | 1.3903 (0.6063) | 0.6482 (0.2372) |

Table 3 Average elasticity tensor measured from juvenile and mature wood

## CONCLUSION

1. A nondestructive testing method has been developed for determining the elasticity tensor of wood or wood-based composites. The approach is based on the wave velocity measurements and theory of polarization transformation of shear wave propagation in a medium.
2. $C_{11}$ and $C_{22}$ show significant differences between juvenile and mature wood.(Figure. 10 and Figure.11)
3. Shear modulus appears to be nearly the same for the juvenile and mature wood. The approximate similarity of $C_{55}$ and $C_{66}$ indicates that wood is at least orthotropic (Figure.14 and Figure.15). Further confirmation of its transversely isotropic property are obtained from the similarity of $C_{12}, C_{13}$ (Figure.16 and Figure.17).
4. Error introduced by modeling tangential direction $(T)$ as a material symmetry should be concerned (Figure.12, Figure.13 and Figure.18).

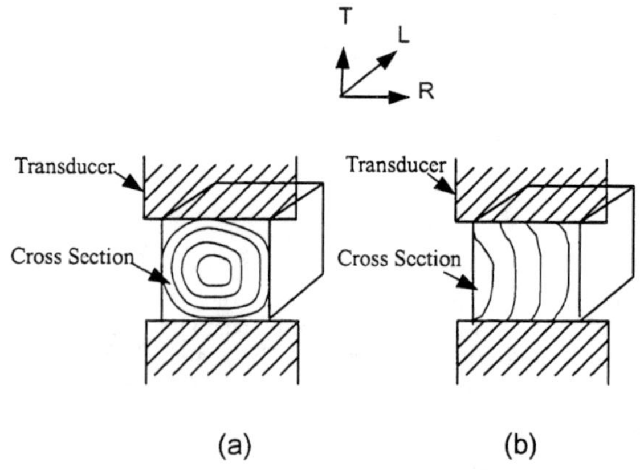


Figure 9. Wood block measured in T direction. (a). Juvenile wood, (b). Mature wood.

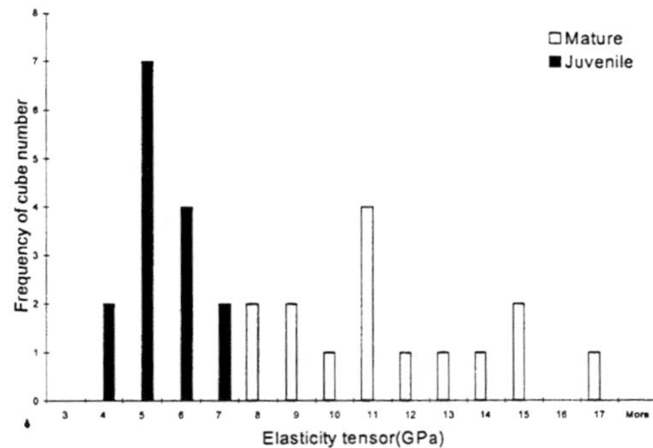


Figure10. Comparison of elasticity tensor C11 between juvenile and mature wood

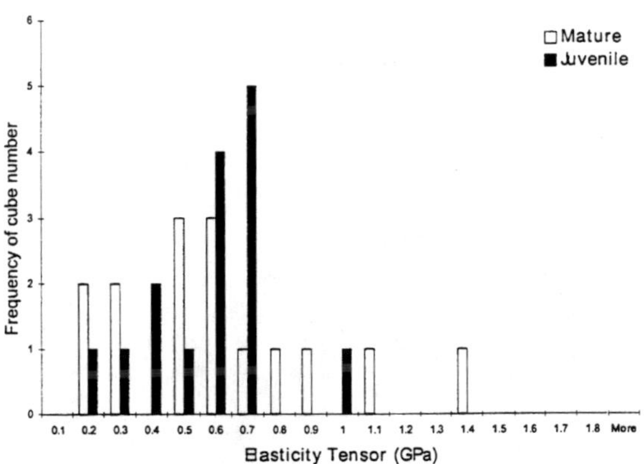


Figure13. Comparison of elasticity tensor C44 between juvenile and mature wood

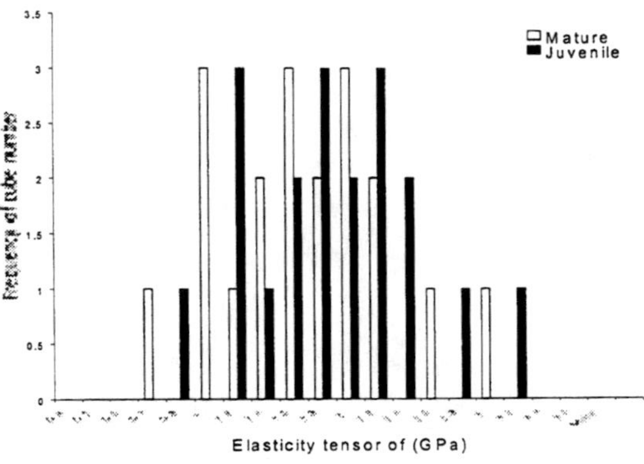


Figure16. Comparison of elasticity tensor C12 between juvenile and mature wood

## REFERENCES


1. Wolfe, R., Year?, "Research Challenges for Structural Use of Small-Diameter Round Timbers," Forest Prod. J., Vol. 50(2), pp. 21-29
2. Bodig, J. and Jayne, 1993, B. A., *Mechanics of Wood and Wood Composites* Krieger Publishing, Malabar, Florida.
3. Peterson, in prep. Tensor transformation
4. Van Buskirk, W. C., and Cowin, S. C. and Carter Jr, R., 1986, "A theory of acoustic measurement of the elastic constants of a general anistropic solid," Journal of Materials Science, Vol. 21, pp. 2759-2762.
5. Norris, A. N., 1988, "On the Acoustic Determination of the Elastic Moduli of Anistropic Solids and Acoustic Conditions for the Existence of Symmetry Planes," Mech. Appl. Math., Vol. 42, pp. 413-426.
6. Aristegui, C., and Baste, S., 1997, "Optimal Recovery of the elasticity Tensor of General Anistropic Materials from Ultrasonic Velocity Data," J. Acoust. Soc. Am. Vol. 101 (2), pp. 813-833
7. Bucur, V., 1995, *Acoustics of Wood,* CRS Press, Inc., N. W.
8. Borgis, F.E., 1955, Phys Rev. 98, 1000 (???)
9. Peterson, M. L., 1997, "A Method for Increased Accuracy of the Measurement of Phase Velocity," Ultrasonics, Vol. 35 (1), pp. 17-29.

Proceedings of
2001 ASME International Mechanical Engineering Congress and Exposition
November 11–16, 2001, New York, NY
NDE-Vol. 21

**IMECE2001/NDE-25814**

# Radiation Beam Modeling and Intelligent Signal Interpretation for Phased Array Ultrasonic Inspection

**Sung-Jin Song**
School of Mechanical Engineering, Sungkyunkwan University
Suwon, Korea

**Hyeon Jae Shin**
Inde System, Suwon, Korea

**Jeong-Rock Kwon**
Korea Gas Safety Corporation, Shihung, Korea

**Abstract**

Flaw characterization with ultrasonic phased array technique involves to key issues, such as obtaining the high quality flaw images and determining the quantitative flaw information (such as location, type and size). This paper deals with these two key issues. For obtaining the high quality images, it is necessary to optimize the parameters of array transducers. To address such a need, a very computationally efficient radiation beam model is developed based on the boundary diffraction wave model, and the 3-D radiation beam fields from array transducers were simulated to investigate their characteristics in detail. From the sectorial images provided by the ultrasonic phased array technique, flaw size can be determined very successfully, if the type of the scatters is identified in advance. For the determination of the type of scatters, an intelligent signal interpretation scheme based on ultrasonic pattern recognition approach is studied, and the variation of features according to the steering angle is found to be a very sensitive feature for this purpose. The performance of the proposed approach is demonstrated with the initial experiments.

## 1. Introduction

Ultrasonic phased array techniques that adopt multi-element array transducers have the well-known capability of providing images of the inside of a test piece in a real-time fashion. The quality of image obtained by this technique is very satisfactory when the test piece contains anomalies with similar acoustic properties, as can be seen in the medical application. In the nondestructive evaluation of structural materials, however, the acoustic properties of flaws are usually not similar to that of the matrix, and then the flaw image obtained by array techniques can be quite different from its real shape. In this situation, optimization of array transducer parameters for obtaining the high quality flaw image and signal interpretation for quantitative flaw characterization become very essential issues.

To provide a suitable remedy to these issues, modeling and intelligent signal interpretation are investigated. Specifically, a model that can calculate full 3-D radiation beam fields from transducers very accurately and computationally effectively is proposed for the optimization of array transducer parameters, and an intelligent signal interpretation scheme based on ultrasonic pattern recognition approach is developed for flaw classification followed by sizing.

## 2. Modeling of Radiation Beam Fields from Phased Array Transducers

### 2.1. Boundary Diffraction Wave Model

The pressure field radiating from a planar piston source can be calculated in a computationally efficient manner by use of the boundary diffraction wave model (BDWM) [1] given by Eq. (1). The BDWM involves a 1-D line integral converted from 2-D surface integral that is known as the Rayleigh-Sommerfeld integral.

$$p(\mathrm{x},\omega)=\rho c v_0\left[\Theta\exp(ikz)-\frac{1}{2\pi}\int_c \frac{\exp(ikr_e)}{r_e}\frac{(\mathrm{n}\times\mathrm{e_r})\cdot ds}{1-(\mathrm{n}\cdot\mathrm{e_r})^2}\right] \quad (1)$$

$$\Theta=\begin{cases}1 & for\ \ y_0\ \ in\ \ S\\ 1/2 & for\ \ y_0\ \ on\ \ C\\ 0 & for\ \ y_0\ \ outside\ \ S\end{cases}$$

where, as shown in Fig. 1, $\mathbf{x}$ is the point in the medium, $\mathbf{y_0}$ is the point obtained by projecting the point $\mathbf{x}$ to the plane of the transducer surface, $\rho_e$ is the radius from $\mathbf{y_0}$ to the transducer edge, $\phi$ is the angle between x-axis and $\rho_e$, $r_e$ is the distance from the point **x** to the point of transducer edge, **n** is the unit normal vector to the transducer, and $\theta$ is the angle between $r_e$ and unit normal vector **n**, $\rho$ is the material density, k is the wave number, c is the wave speed of the material, $v_0$ is the spatially uniform velocity of the transducer surface, and $\mathbf{e}_r$ is the unit vector.

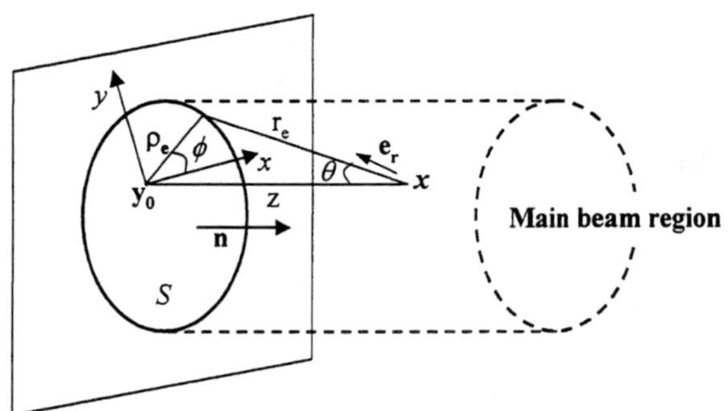


**Fig. 1. Geometrical terms when pressure is evaluated at a point x.**

In fact, the BDWM is a quite general model so that it can be used to calculate beam fields from various transducers including planar transducers with arbitrary cross-sections. Recently, Song and Kim[2] have shown the effectiveness of this model to the calculation of the radiation fields from rectangular piston sources. As a natural extension of their work, the radiation field from a phased array transducer can be obtained by the superposition of the radiation fields from individual array elements with proper phase delays.

**2.2. Simulation**

Configuration of an array transducer used in this simulation study is shown in Fig. 2. Center frequency of the transducer is chosen to be 7.5 MHz and hence the wavelength, $\lambda$, in steel is about 0.79 mm. To avoid grating lobe, the pitch, *p*, is set as 0.3 mm that is less than half of the wavelength. The inter element gap is 0.1 mm that is sufficient to suppress the cross-talk between neighboring elements. The lateral dimension, *D*, of the transducer is 19.1 mm, while the element length in the elevation direction is 6 mm. Thus, the near field length, $D^2/(4\lambda)$, of the transducer becomes about 115 mm.

Focusing and steering of the radiation beam were done in the lateral plane that passes through the center of the array transducer. Electronic delay times are applied for the focusing and steering at the 10 points (F1, F2, ..., F10) shown in Fig 3. All of the points are located in the near field of the transducer and evenly spaced with the spacing of 10 mm. In the case of F10, for example, ultrasonic beam is electronically steered to 45 degrees and focused at the 50 mm in axial depth.

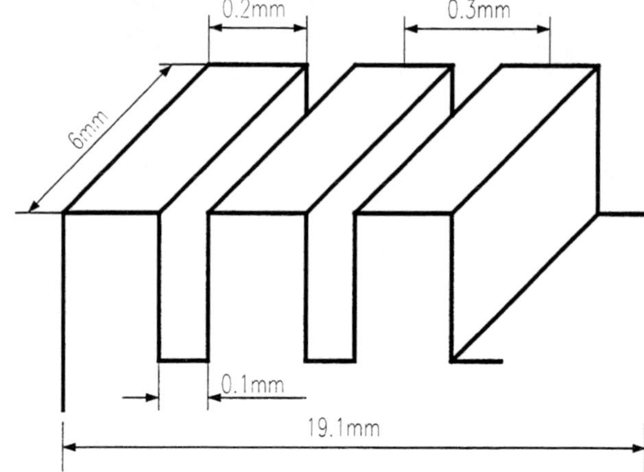


**Fig. 2. Geometry of an array transducer**

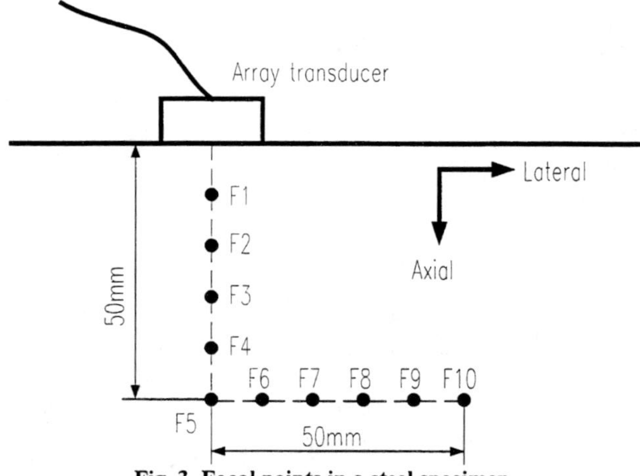


**Fig. 3. Focal points in a steel specimen.**

Fig. 4 shows an example of radiation fields from the transducer when the beam is focused at (a) infinity without steering, (b) F2, (c) F5, and (d) F10 by electronic time delay. The beam is well focused within near field length without grating lobe. As the focal point comes closer to the transducer, the beam beyond the focal point spreads out very quickly.

Fig. 5 (a) and (b) show the on-axis pressure field when the beam is focused at F3 and F5, respectively. The actual focal points, which have the maximum on-axis amplitudes, appear ahead of the electronically aimed focal points, and the deviation gets bigger as the focal point is deeper. In fact, when the electronic focal point is infinity, the array transducer focuses naturally at the near field length. Fig. 6 shows the comparison of the actual focal depths to the aimed focal depths together with the lateral sizes of the actual focal points.

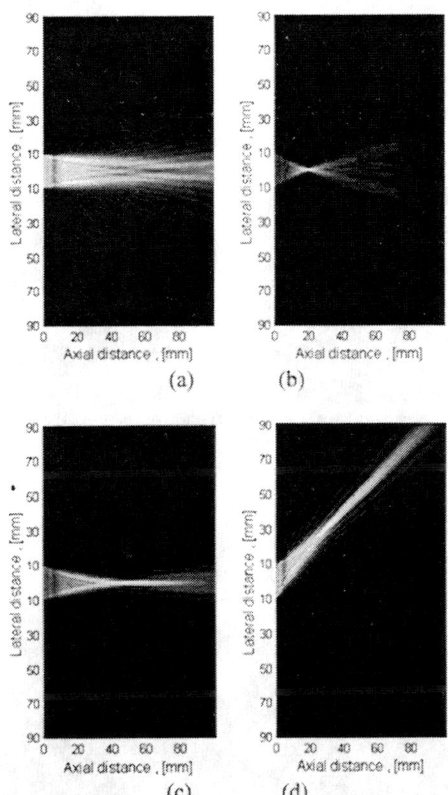

Fig. 4. **Radiation beam fields from the array transducer when the beam is focused at (a) infinity without steering, (b) F2, (c) F5, and (d) F10 by electronic delay time.**

### 2.3. Fabricated Array Transducers and Experimental Results

The simulation studies confirmed that there would be no grating lobes when the ultrasonic beam is steered up to 45 degrees. Therefore, the array transducer shown in Fig. 7 (a) was fabricated with the geometrical specifications shown in Fig. 2. The number of element is 64 and the center frequency is 7.5 MHz. This transducer can be used for the sectorial scan in which the view angle is less than 45 degrees in one direction. Fig. 7 (b) shows the array transducer having 128 elements. The pitch of this array transducer is 0.6 mm, the inter element gap is 0.1 mm, and the total lateral dimension is 76.5mm. In this geometrical specification, there are no grating lobes if ultrasonic beam is not steered, so that the 128-element array transducer is used for linear scan without steering.

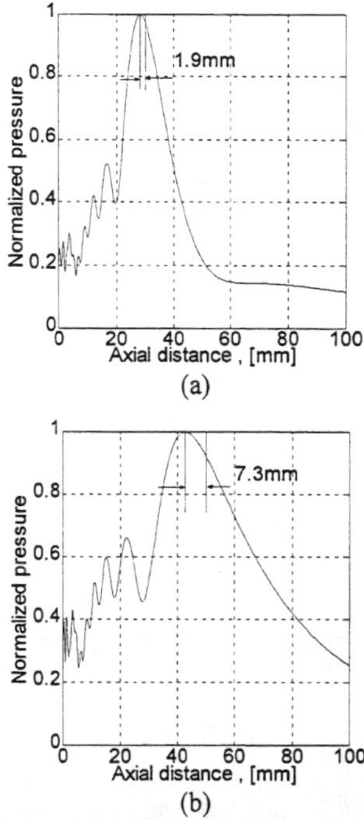

Fig. 5. **On-axis pressure fields for on-axis when the beam is focused at (a) F3 and (b) F5.**

Fig. 8 shows electronically scanned ultrasonic images by using the fabricated array transducers and the phased array system that has been developed by authors and their coworkers[3]. The specimens are made with carbon steel and have side drilled holes. Fig. 8 (a) shows the sectorial scan image obtained by using the 64-element array transducer shown in Fig. 7 (a). Fig. 8 (b) shows the linear scan image obtained by using the 128-element array transducer shown in Fig. 7 (b). In both results, the defect are well detected and easily located.

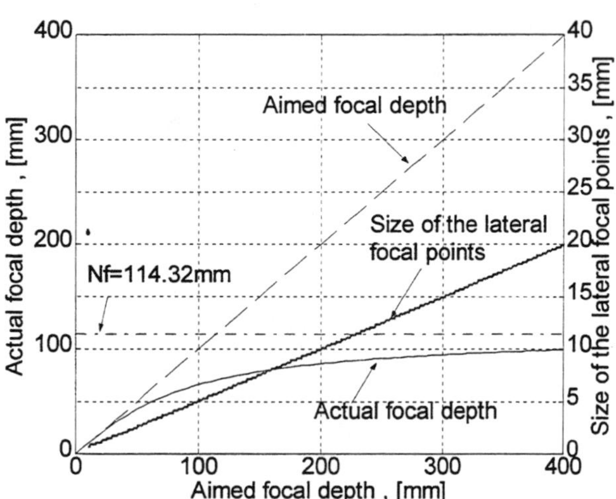


**Fig. 6. Comparison of the actual focal depth to the aimed focal depth together with lateral size of the actual focal points.**

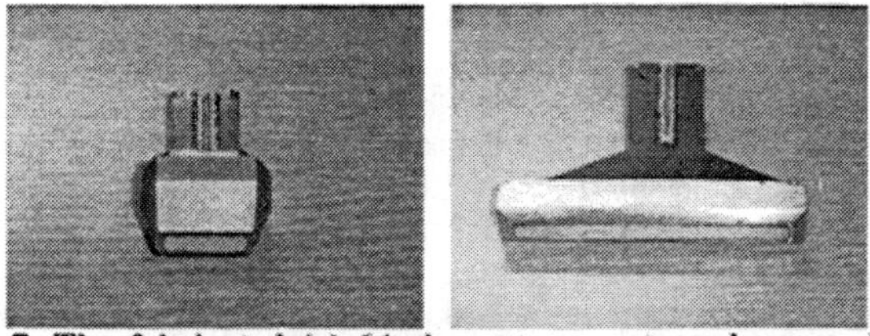

**Fig. 7. The fabricated (a) 64 element array transducer and (b) 128 element array transducer.**

## 3. Intelligent Signal Interpretation for Quantitative Flaw Characterization

Since the ultrasonic phased array technique provides two-dimensional images of the inside of steel structures it becomes relatively easy to determine the exact location of flaws in comparison with the classical A-scan inspection. Even with the phased array technique, however, it is still difficult to decide the real shape of flaws from their images. For example, the reflection from a side-drilled hole is hard to distinguish from the diffraction from a crack tip in its image. Fig. 9 shows the two images obtained from two different types of flaws. Fig 9 (a) was obtained from a specimen with the horizontal crack of 9.6 mm long showing the two diffractions from the left tip (denoted by 'A') and the right tip (denoted by 'B') of the crack. Fig 9 (b) was obtained from a specimen having two side-drilled holes horizontally spaced out 9.6mm apart showing the reflections from the left and the right holes, denoted by 'C' and 'D', respectively. This proclaims the difficulty in the discrimination of cracks from holes from their images.

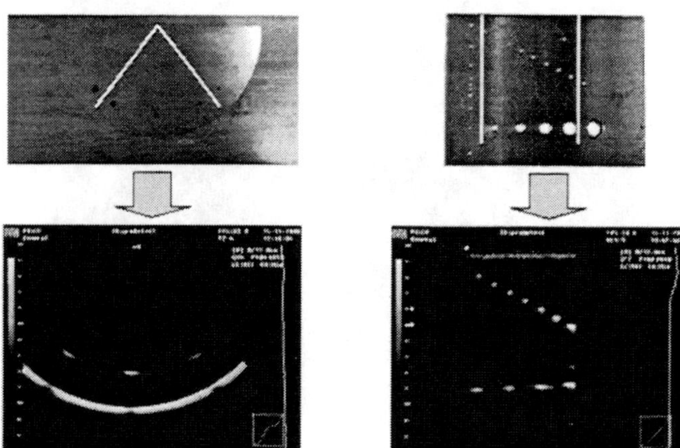

**Fig. 8. Electronically scanned ultrasonic images by using (a) 64 element array transducer and (b) 128 element array transducer.**

To take care of this difficulty, an intelligent signal interpretation methodology that can automatically classify flaws into three categories (side-drilled holes, horizontal cracks, and vertical cracks) as shown in Fig. 10 is proposed based on the ultrasonic pattern recognition approach developed by Song and Kim (2000)[4].

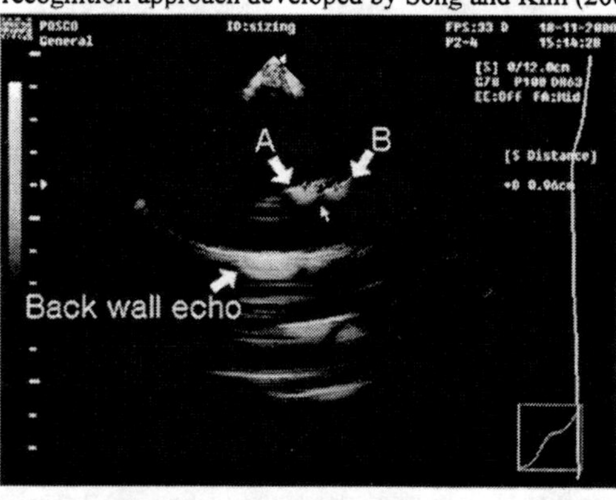


(a)

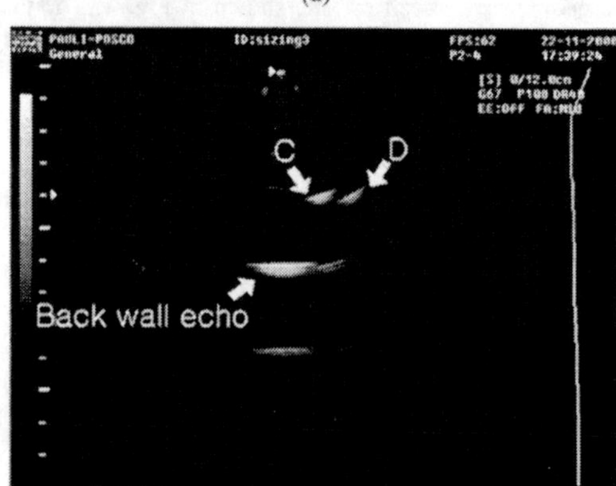

(b)

**Fig. 9. Electronically scanned sectorial images obtained from (a) a carbon steel specimen with 9.6mm long horizontal crack and (b) a carbon steel specimen with two side drilled holes that are horizontally spaced out 9.6mm apart.**

The phased array ultrasonic inspection system that has been developed by the authors and their coworkers[3] has the capability of providing the RF signals for all scan lines. Taking advantage of this great capability, a set of 18 ultrasonic features (listed in Table 1) was extracted from the RF signals captured from the flaw under investigation. This set of features was extracted from six RF signals captured at six different steering angles of 0°, 5°, 10°, 15°, 20°, and 25°. Then, the variations of the 18 feature values (in terms of the slopes) with the steering angle were investigated for three types of flaws.

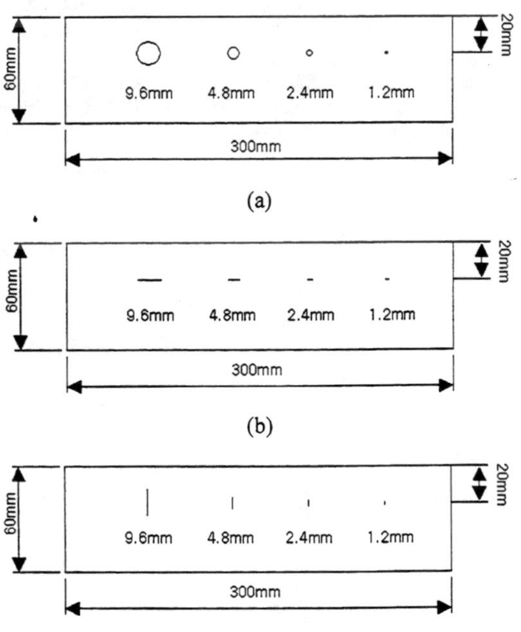


(c)

**Fig. 10. Various sizes of holes and horizontal and vertical cracks made in carbon steel blocks 40 mm thick for flaw classification experiments. The diameters of the holes and lengths of the cracks are indicated under the each flaw.**

Fig. 11 shows an example of the results of this analysis for two features, F1 (peak amplitude) and F5 (energy of the 1st group), those were turned out to be useful for flaw classification. Figs. 11 (a) and (b) show the variations of the two feature (F1 and F5) values for three types of flaws (the horizontal crack, the side drilled hole, and the vertical crack) with the size of 2.4 mm. The both feature values show similar trends not only for the changes of the steering angle but also for the flaw size. As the steering angles were increased, the feature values were rapidly decreased (with a negative slope) for horizontal cracks and were rapidly increased (with a positive slope) for vertical cracks, while the changes of the feature values were negligible (with a negligible slope) for side-drilled holes.

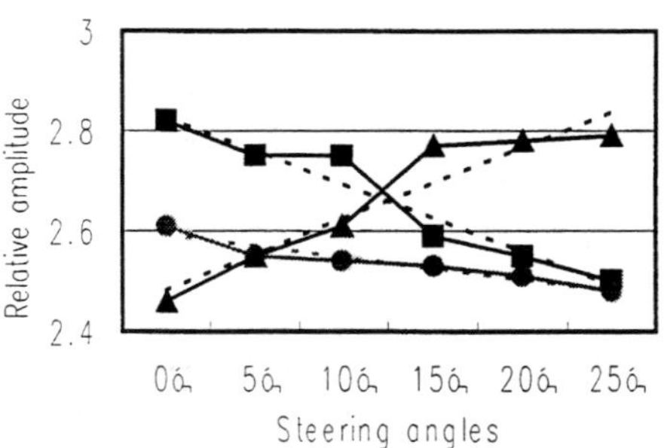


(a) The variation of the feature, Fl

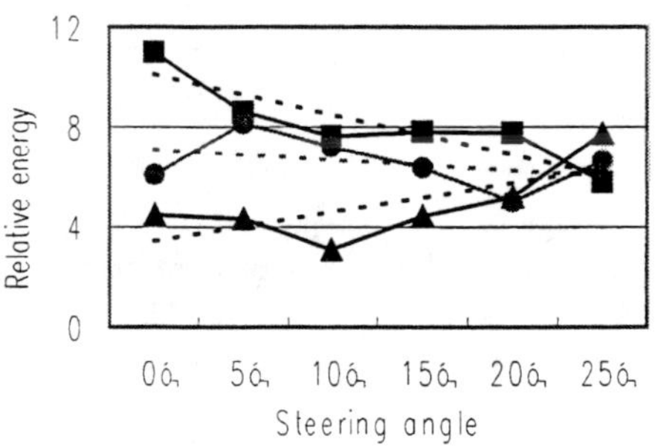


(b) The variation of the feature, F5

**Fig. 11. The variations of (a) the feature F1 (peak amplitude) and (b) the feature F5 (energy of the 1st group) for the various steering angles. The diameters of side drilled holes and the lengths of the cracks are 2.4mm (■ : horizontal crack , •: side drilled hole , ▲: vertical crack)**

Fig. 12 provides the information on the slopes of the variation of two features (F1 and F5) according to the change of the steering angle for four different flaw sizes. The horizontal cracks have negative slopes for all flaw sizes, and the vertical cracks have positive ones. The side-drilled holes, however, have negligible slopes. This result demonstrates the effectiveness of the slope of feature variation for flaw classification.

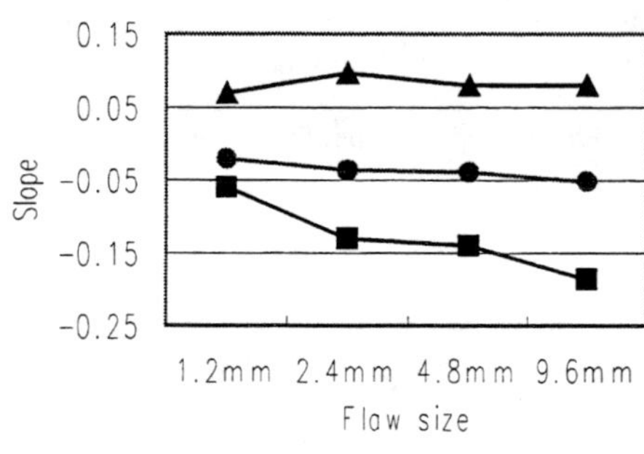


(a)

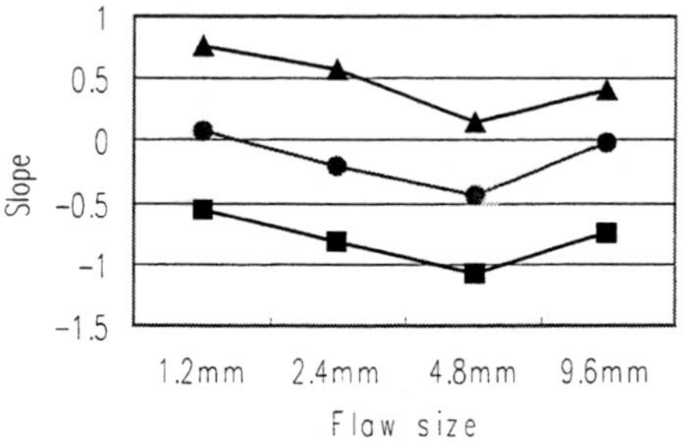

(b)

**Fig. 12. The signal of the variations of (c) F1 and (2) F5 according to the change of the steering angle for four different flaw sizes. (■ : horizontal crack , •: side drilled hole , ▲: vertical crack)**

## 4. Concluding Remarks

Obtaining the high quality flaw images and determining the quantitative flaw information (such as location, type and size) are two key issues in the flaw characterization using ultrasonic phased array techniques. This paper addressed both issues.

For the optimization of parameters of array transducers, which is essential for obtaining the high quality flaw images, a very computationally efficient radiation beam model was developed based on the boundary diffraction wave model. The 3-D radiation beam fields from array transducers were simulated and investigated in detail. Then, using this beam model, array transducers were fabricated and validated their performance by the image capturing experiments.

From the flaw images provided by the ultrasonic phased array inspection technique, flaw size can be determined very successfully, if the type of the scatters is identified in advance. The identification of the scatter, however, is not trivial. To address such a need, an intelligent signal interpretation scheme based on ultrasonic pattern recognition approach was studied in this work. From this endeavor, the variation of features according to the steering angle was found to be very effective for the determination of the type of scatters, and the performance of this approach was demonstrated with the initial experiments.

## Reference

1. Schmerr, L. W. Jr., Sedov, A., Lerch, T. P., 1997, "A boundary diffraction wave model for a spherically focused ultrasonic transducer" The Journal of the Acoustical Society of America. Vol. 101, No. 3, pp. 1269-1277.
2. Song, S. J., Kim, H. J., 2000, "Modeling of radiation beams from ultrasonic transducers in a single medium", Journal of the Korean Society for Nondestructive Testing, Vol. 20, No. 2, pp. 91-101.
3. Hyeon Jae Shin, Song, S. J., You Hyun Jang, 2000, "Nondestructive Inspection of Steel Structures Using Phased Array Ultrasonic Technique", Journal of the Korean Society for Nondestructive Testing, Vol. 20, No. 6, pp. 538-544.
4. S. J. Song and H. J. Kim, 2000, "An Intelligent System Approach to Real-Time Ultrasonic Flaw Classification in Weldments", JSME International Journal, Series C, Vol. 43, No. 1, pp. 60-72.

Proceedings of
2001 ASME International Mechanical Engineering Congress and Exposition
November 11–16, 2001, New York, NY
NDE-Vol. 21

IMECE2001/NDE-25815

# Inspection Of Fastener Holes Using Ultrasonic Phased Arrays

Pamela G. Herzog
Air Force NDI Office

Vincent Lupien
Acoustic Ideas Inc

James T. Miller
Lockheed

John J. Selman,
Lockheed

Michael Moles
R/D Tech

## ABSTRACT

An Air Force Aging Aircraft Project was initiated to identify a suitable replacement for the Autoscan. The Autoscan is a unique piece of equipment that was originally identified for specific inspections for detection of first-layer, faying surface fatigue cracks (.030" and larger) around fastener holes beneath the fastener heads, without removal of the fastener. The currently inspected parent material ranges from .125 to .3 inches thick, 2024T3 or 7075-T73 aluminum. Potential also exists for replacing other aging inspection equipment such as the Rotoscan. A secondary objective was to evaluate the feasibility of expanding the inspection capabilities to detect corrosion as well as cracks.

A Trade Study was conducted initially to consider existing technologies available, trade-offs, and technology insertion in order to meet the required performance parameters. The trade study showed Phased Array Ultrasonics to have the greatest potential, so it was chosen as the inspection method to pursue. Using phased arrays, a novel inspection technique for rapidly and reliably inspecting the area around fastener holes for cracks and corrosion has been developed *with no moving parts.* Specially designed probes are used for the aircraft inspections.

This design consists of a three-dimensional matrix of 504 ultrasonic elements on a cone that encircles the fastener head. The two-dimensional arrangement of elements permits deflection of the ultrasonic beam in three dimensions. Full circumferential scans are performed by programming the phased array focal laws to scan 360° of the fastener holes, using a combination of the following scan patterns: pulse-echo at 45° incident on the crack, pulse-echo at 90°, pitch-catch, plus local scanning. This capability allows flexible coverage of the fastener hole and surrounding area, again with no moving parts. Additionally, the beam deflection capability means that one probe is adaptable to a wide range of fastener diameters and skin thickness.

Several conical sub-arrays were built to evaluate the feasibility of the concept experimentally. The experimental results along with numerical modeling were used to determine optimal values for inner and outer radii of the cone, angle of the cone, number of elements and arrangement of the elements. A complete prototype conical array was subsequently built.

The final portion of this project includes developing the specific inspection procedures, and performing a Probability of Detection study (POD) developed by the FAA's Airworthiness Assurance Nondestructive Tested Validation Center at Sandia National Laboratory.

## BACKGROUND

In the early 1980's, a fastener inspection system known as the Autoscan was designed and built to perform full circumferential scans around aircraft fasteners to search for faying surface corner fatigue cracks 0.030" and larger at the edge of fastener holes in the top layer of aircraft skin structures without removal of the fastener. It was designed to accommodate a variation in top layer skin thickness from 0.1" to 0.5" and in fastener hole diameter from 3/16" to 3/8". The unit uses a pair of mechanically rotating ultrasonic probes for the inspection, and eddy current sensors to aid in centering the unit on the fastener. The wide range in skin thickness and fastener diameter is accommodated by mechanical adjustments in the position of the probes.

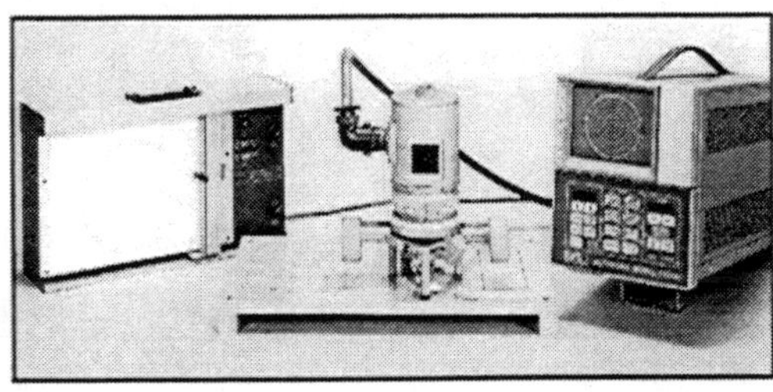

Original Autoscan System

The Autoscan has now reached obsolescence and is no longer maintainable, but the faying surface fastener inspections are still needed to ensure safety of the aging fleet. An Air Force program was initiated to find a suitable replacement, starting with a trade study to review currently available technologies. The trade study was performed seeking latest state of the art technologies that could apply to multiple aircraft programs.

The trade study revealed ultrasonics to be the method of choice, as it was during the design of the original Autoscan. With the advent of new approaches in ultrasonics however, there is now considerably more variety.

Before proceeding to an explanation of the selected approach, we will review some of the areas of the original system where improvement was sought in considering candidates for the replacement system. The foremost limitations in the Autoscan were its weight of approximately 20 pounds and its slow scanning time. The weight made manipulation of the instrument difficult for an operator and required the use of an external support device for under-wing inspections. Along with reductions in weight and scanning time, reductions in size were sought in the new scanning head, for the same ergonomic reasons.

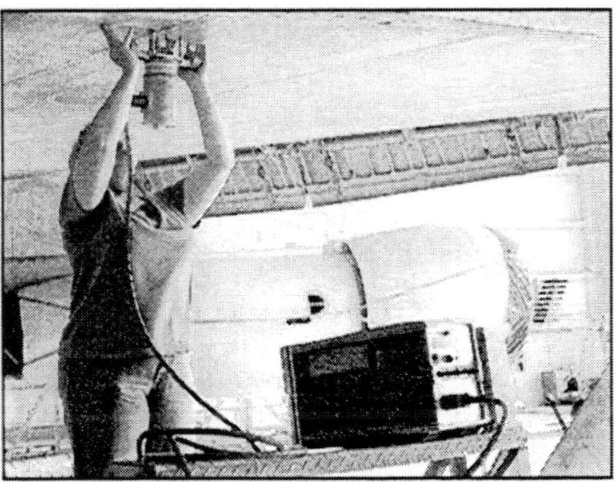

Autoscan Wing Fastener Inspection

Given that the redesign was being performed 20 years after the Autoscan, it was expected that technological development could increase functionality while improving the ergonomics. In particular, the ability to inspect standard shaft cracks, countersink cracks, corrosion mapping and skin thickness measurements along with faying surface cracks is desirable.

Phased array approaches offered the possibility of reducing weight and size while increasing performance.

## PHASED ARRAY ULTRASONICS

Phased arrays are an electronic method of generating and receiving ultrasound, which permits beam steering, beam scanning and beam focusing. In a phased array, each element is individually wired to a pulser/ multiplexer and time delay circuitry. A selected series of these elements is fired with pre-determined delays to build up multiple wavefronts in the component. The wavefronts interact, generating zones of constructive and destructive interference that yield the desired beam angle and focal spot position. By adjusting which elements are active and the set of time delays, the beams can be electronically steered and scanned to give complex inspection patterns. Using an array, *hundreds* of different conventional probes with varying angles and focal lengths can be simulated. An example of an angle beam formed from a linear array is shown below.

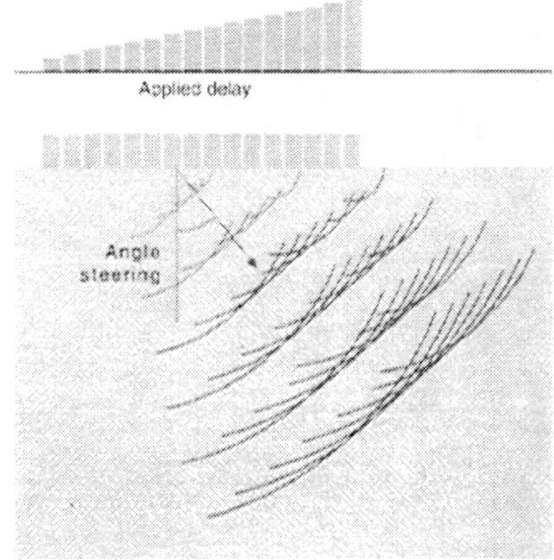


Forming an Angle with A Linear Array

**Basic Set-Up Procedure**

The operator selects the refracted angle and focal spot from a set-up table, and a computer calculates the appropriate elements to pulse and the time delays required. Once calculated, the operator saves these Focal Laws as a set-up file for rapid set-ups.

The computer sends these Focal Laws to the phased array instrument, which performs the pulsing and receiving.

The collected waveform data is sent to a computer for processing, which displays the information in the traditional A-, B-, C- and D-scans (also known as "top, side, end" views). The advantages of phased arrays are their flexibility, high speed, small footprint and rapid set-ups. Phased arrays are useful in a wide variety of industrial applications.

**A New Approach**

Using phased arrays, a novel inspection technique for rapidly and reliably inspecting the area around fastener holes for cracks was developed *with no moving parts.* Special probes using conical array of ultrasonic crystals have been designed for the aircraft fastener inspections. The probe developed is a three-dimensional conical array of ultrasonic elements laid out as shown below.

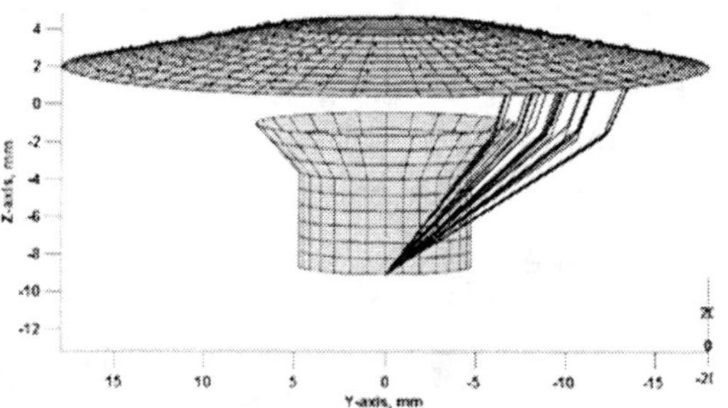


Probe Side View w/Fastener

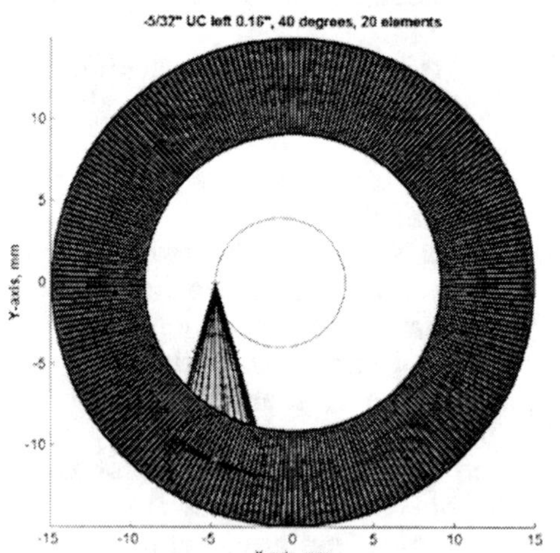


Probe Top View w/Fastener

## Feasibility Study

A Feasibility Study was initiated to determine if a practical probe could be produced reliably and economically. A primary concern was to minimize the presence of grating lobes. Grating lobes are known to appear when empty space exists between adjacent elements, and can generate misleading false signals if not sufficiently suppressed. To eliminate the empty space in the conical matrix design would have required over 4000 ultrasonic elements, which was neither technically feasible nor economical.

The challenge, from an ultrasonic point of view, was to select the largest elements possible that still have enough deflection power to generate the desired beams for all the desired fastener diameters and skin thicknesses, using the smallest number of elements possible while still maintaining acceptably low grating lobes. These elements will be distributed over a conical shape whose geometry must provide enough aperture to generate a 0.030" focal spot for all inspection scenarios, including centering operations and corrosion detection.

The probe was designed using computer simulation to determine optimal values for:

- Number of elements
- Angle of cone
- Inner and outer radii of cone
- Size and shape of elements
- Waterpath

Computer modeling was used to simulate results for a set of sample cases representing extremes in the performance envelope. Each sample case included four sub-cases corresponding to the array being off-center by the centering tolerance towards the left, right, top and bottom. For each focal law, the number of active elements required was determined. The shape and size of the probe was initially determined from the simulation results.

The modeled fields demonstrated that the array could be used to steer the focal spot to the all the desired depths and radial positions. The fields also indicated the effect of the large inter-element spacing (primarily grating lobes). The goal was to keep the maximum grating lobes at 20 dB below or more below the amplitude of the focal spot. Known strategies to reduce peak grating lobe levels include randomizing the positions of the elements and minimizing the size of the cone further so that the elements are closer together.

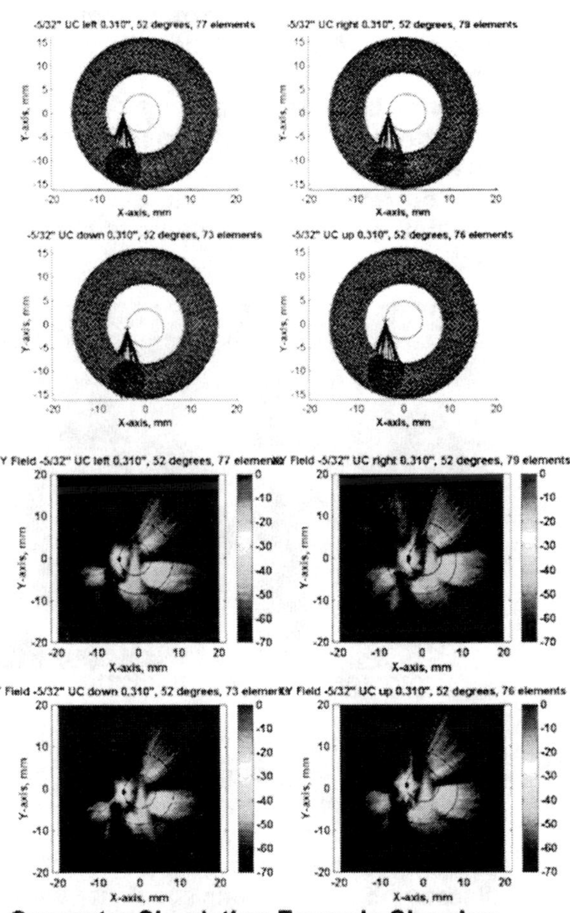


**Computer Simulation Example Showing Array Off-Center in 4 Directions**

Several conical sub-arrays (Partial probes) were constructed to validate the design experimentally and to verify manufacturing feasibility.

The Feasibility results indicated that detection capability was high, potentially above the requirement of 0.030" on the first layer faying surface cracking.

### Prototype Probe Design

The fastener inspection system uses a conical matrix array for self-centering and for scanning the fastener holes. The system consists of a 504-element array with instrumentation capable of addressing up to 512 possible elements with 64 multiplexed pulsers.

This design permits three-dimensional beam deflection. The beam can be steered and focused to suit the fastener diameter and skin thickness variations. The 360° array permits full coverage of the fastener without any probe motion. The scan around the fastener head takes place electronically through a rapid multiplexing operation. The frequency of the probe is 10 MHz. The current array design uses "quasi-random" element positioning to minimize grating lobes.

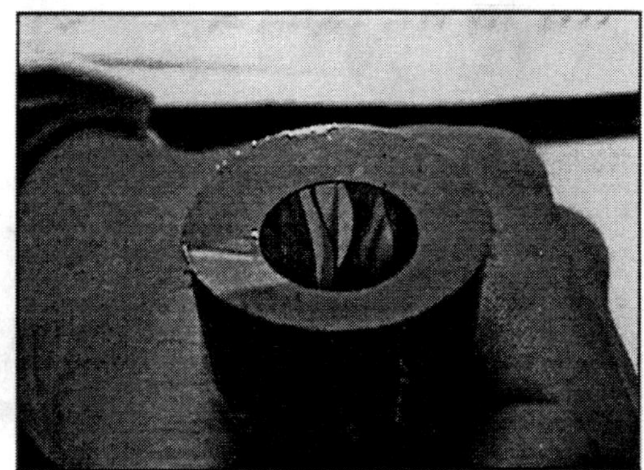

Prototype Conical Matrix Probe

### Water Coupling

The coupling mechanism is a water filled membrane with small holes allowing water to seep through at the rate determined necessary. A photo of the disassembled coupling boot is shown below.

Photo of Coupling Boot w/Mold

For centering, the operator will manually set the probe within +/- 0.060" of true center. A spring-loaded mechanical "pointer" is incorporated to assist the operator with this initial step. The pointing device is shown in the partial assembly photo shown below. Also indicated are three hard rubber feet that position the array the proper distance over the fastener head. This design permits inspection of protruding head fasteners as well as flush head fasteners.

Mechanical Pointer and Feet

Once centered within +/- 0.060" of true center, the system will aid the operator to center to within 0.030". Once this centering accuracy is achieved, the system will:

- Indicate satisfactory positioning,
- Measure the hole location,
- Use a look-up table to download the necessary Focal Laws, and
- Perform the scan.

Subsequently, the array will verify the position of the probe around the fastener to ensure that no probe movement occurred during the scan. The entire sequence is completed in a matter of seconds. The data will be displayed as a C-scan map with the fastener hole position indicated, and any defects shown in color; however in principle, almost any display is possible.

By transmitting and receiving with the same set of elements, pulse-echo measurements can be made. By simultaneously receiving with a set of elements on the opposite side of the array, a pitch-catch measurement can be made at the same time.

### Probe Holder

The phased array probe and water coupling is housed in a plastic holder similar in size and appearance to a hair dryer as shown below. The encasement for the probe was designed to make it easily hand held.

Probe Holder Assembly

An electronics module is housed in the probe holder for local control functions. Switches are located at the rear of the housing assembly for initiating the centering assist process, initiating data collection, and accepting/storing data.

### System Controller

The system controller is a customized FOCUS 64/512 phased array system controller developed by R/D Tech Inc. A photo of the system controller is shown below.

FOCUS 64/512 System Controller

### Summary

The objective of this project was to develop a lightweight, hand-held device for rapidly and reliably inspecting the area around fastener holes for faying surface cracks, standard shaft cracks and countersink cracks, without removal of the fastener.

Using phased arrays, a new inspection technique for rapidly and reliably inspecting the area around fastener holes for cracks has been developed *with no moving parts.* Special probes using conical array of ultrasonic crystals have been designed for the aircraft fastener inspections. A prototype probe with 504 elements was built, and integrated into a 64/512 Focus system. Further capability is currently being developed. This additional capability includes corrosion mapping, pitch-catch inspections, skin thickness measurements, detection of other defects such as exfoliation and shaft cracking. The final portion of this project includes developing specific inspection procedures and performing Probability of Detection (POD) tests to verify the capabilities. The finalized product with full defect characterization is scheduled for delivery around December 2001.

Proceedings of
2001 ASME International Mechanical Engineering Congress and Exposition
November 11–16, 2001, New York, NY
NDE-Vol. 21

IMECE2001/NDE-25816

# EFFECT OF CORROSION DAMAGE ON FATIGUE BEHAVIOR OF AL 2024

Mohammad Khobaib[1]

Michael S. Donley[2]

[1]University of Dayton Research Institute, Dayton, OH
[2]Air Force Research Laboratories, Wright-Patterson AFB, OH

## ABSTRACT

Fatigue crack nucleation from corrosion damage is drawing considerable attention in the aging aircraft research community. Localized corrosion, such as corrosion pits, is a potential stress raiser and provides active sites for fatigue crack initiation. The detailed knowledge of the metrics of early stages of corrosion damage and transition to fatigue cracks is critical to the accurate life prediction of the structure. High strength aluminum alloys are known to be very sensitive to orientation and this could play a critical role in both pitting and the resulting fatigue damage.

Fatigue experiments on dog bone specimens of Al 2024-T3 with known pit morphology were conducted to investigate the crack initiation and growth behaviors of cracks nucleating from pits. The corrosion pits were grown electrochemically in a controlled fashion and samples were prepared with pit depth varying from 3 to nearly 130 μm. These pits were characterized using white light interferometer; surface parameters such as average roughness and morphology were determined before fatigue testing. Specimens were aligned in three different directions, L-T, T-L and L-S to investigate the effect of orientation on pitting and fatigue characteristics.

The fatigue lives were observed to decrease with increasing pit depth. The fractured surface was analyzed after the fatigue test and stress intensity calculations were made based on these observations. A linear relationship between stress intensity factor and critical pit depth was observed. The crack growth behavior of the crack nucleating from these pits was typical of a short crack. The results provide the details of the critical pit size that can lead to fatigue crack initiation as well as supply information about the growth behavior of cracks nucleating from corrosion pits. An understanding of the role of pitting in crack initiation aids in extending the fatigue life of the aging aircraft.

## 1 INTRODUCTION

Corrosion of high strength aluminum alloy structures poses a serious problem in extending the life of aging aircraft fleet. The United States military has a large fleet of aging aircraft, which is expected to grow further due to budgetary restriction. Most of these aircraft will be in operation for longer than they were originally designed to be [1]. Consequently, the Air Force is concerned about the growing number of aging aircraft that require extensive maintenance. The older these aircraft get, the maintenance of their structural integrity becomes increasingly critical for life sustainment. Hence increased attention is required towards structural damage modeling, repair and life prediction.

Aircraft structures experience a variety of environmental influences even under normal service conditions and some of these are synergistic while others are cumulative. Some of the main causes of failure of aging aircraft is the cumulative corrosion damage, fatigue cracking and their synergistic interactions. These problems are bound to worsen with the continued service use of these aircraft. The Department of Defense is addressing these issues with research initiatives related to corrosion problem facing aging aircraft, structural testing and developing analytical tools for qualifying the effects of corrosion on structural integrity [2].

The corrosion occurs due to a number of mechanisms, and several different forms of corrosion have been observed in aircraft structures subjected to differing field conditions. General corrosion, pitting and crevice corrosion, intergranular corrosion and galvanic corrosion are among the most common forms observed in aging aircraft structures. The high strength 2000 and 7000 series aluminum alloys, widely used in airframe structure possess heterogeneous microstructures are highly susceptible to localized corrosion such as pitting. Furthermore, the influence of the microstructure and orientation, both on the initiation of pitting and its transition to active notch has hardly been addressed.

Pitting is the type of corrosion most commonly observed in these high strength aluminum alloys. Although it is well known that the conjoint action of corrosion and stress leads to accelerated loss in the structural integrity of aircraft, the critical stage of damage where, under stress, a corroded area leads to crack initiation is not understood at all. One example of this mechanism is the transition of pitting corrosion to active notch. The corrosion is recognized as one of the main nucleation sites for fatigue crack formation in a wide range of steels and high strength aluminum alloys [3-10].

The most significant problems are usually associated with airframe components, such as rivets, fasteners and joints. To date, the most practical solution for inspection of corrosion damage in these areas is to completely strip the paint from the plane. This method is naturally quite expensive, and time consuming and sometimes no corrosion are found. There is no predictive tool to assess the life of structure from any corrosion metrics. Quantitative measures such as pit dimensions and surface roughness as corrosion metrics have been used along with crack growth tools in estimating fatigue life of components subjected to corrosion [11]. Other studies have focussed on measuring and applying a large number of candidate metrics for quantifying corrosion in aging KC-135 [12]. However, these results can be used to predict specific cases, because the very sensitive issue of orientation was ignored in all these investigations.

Pitting is thus a fundamental form of corrosion that occurs in aircraft structural materials. The above descriptions point to a need for the thorough investigation of pitting corrosion and corrosion-fatigue interactions and orientation effects as they apply to structural integrity and life prediction. In addition, as aircraft age and corrosion accumulates, pits at numerous locations may become sources for multiple-site damage and widespread fatigue damage. It is well known that the conjoint action of corrosion and fatigue leads to accelerated loss in the structural integrity of aircraft

This study was initiated under the sponsorship of Department of Defense, to quantitatively investigate the role of corrosion damage on the fatigue behavior of high strength aluminum alloys. It must be emphasized that the pitting effect on fatigue is not yet well understood. A large number of laboratory investigations have studied the effects of prior corrosion and the presence of a corrosive environment on fatigue crack initiation and growth of high strength aluminum alloys [13,14]. These studies, however, have not addressed the early stages of corrosion and its transition to fatigue cracking in structural materials. The information presented in this study about short-term as well long-term effects of corrosion is vital to understanding the early stages of damage evolution in aircraft structure whose operational mechanical and environmental spectra are not adequately represented in laboratory experiments. Nakajima and Tokaji [15] have found that a crack emanates from a pit during simultaneous corrosion fatigue only if the pit grows to the level when the stress intensity factor reaches a certain threshold value. Phillips and Neuman [16] demonstrated that the fracture mechanics of long crack fatigue growth in Al 2024-T3 alloy leads to substantial exaggeration of the fatigue life for a very small initial size. They established the importance of the short crack analysis and experimentally found Paris constants for short and long crack growth in Al 2024-T3 alloy.

In this study controlled pits were grown in standard dog-bone fatigue samples fabricated with tensile axes parallel to L-T, T-L and L-S orientation. These specimens were then tested under fatigue condition in laboratory air. The high strength aluminum alloys are known to be orientation dependent [17,18]. The primary objective of this study is to develop a relationship between a pitting parameter and a structural integrity parameter. The overall objective is to formulate a life prediction model based on pitting parameters obtained from nondestructive evaluation tool used in the service and expected fatigue spectra of the aircraft. Initially a method was developed to produce controlled pitting, so that pits of varying sizes, morphologies and distributions could be created on fatigue samples fabricated from Al 2024-T3. A number of Al 2024-T3 dog-bone fatigue samples were electrochemically pitted. The pit morphology was characterized using optical microscopy, white light interference microscopy (WLIM), and scanning electron microscopy (SEM). All samples were fatigued under the same conditions. The fracture surface of all the specimens after failure was characterized using optical microscopy and SEM and pit dimensions and crack geometry and parameters were obtained. The stress intensity factor calculations were made using Neuman-Raju approximation [19] and, finally the crack growth behavior was obtained through software package, CRACKS [20].

## 2 EXPERIMENTAL

A large number of sets of dog–bone fatigue samples of Al 2024-T3 were machined with L-T, T-L and L-S orientation. The dimensions for the fatigue specimens are shown in Figure 1. Each sample was cleaned and prepared at a time. The gage section of the sample was sanded up to 600 grit SiC paper. The sample was then rinsed with distilled water and dried with a stream of air. Most of the specimen's surface was covered with a waterproof tape. A small portion in the center of the gage section was initially left exposed as shown in Figure 1, and later covered with a thin coat of micro-stop. Two fine holes were created in the microstop coating after the film had dried. This was intentionally done to provide desired site/sites for pit initiation. The sample thus prepared was then glued into an electrolytic cell for electrochemically producing pit/pits on the surface.

A solution of 0.1M NaCl was used as the pitting electrolyte. A Saturated Calomel Electrode (SCE) was used as the reference while platinum meshed disc served as the counter electrode. The pitting was produced in the area of the exposed gage section using a potentiostatic scan. An initial potential of –450 mV (SCE) was applied for 10 minutes, and then the potential was lowered to –520 mV (SCE) for the remainder of the scan. Various scan times were used to obtain pits of varying depth. The sample was promptly removed from the cell, after the potential scan, cleaned with running distilled water and dried in a stream of air. The pit morphology and depth measurements were determined using optical microscopy, SEM and WLIM. The samples were then fatigued within 24 hours of being removed from the electrolytic cell. Fatigue tests for L-T and T-L were conducted on a single machine under the same conditions at a stress ratio, $R = 0.1$, maximum stress = 256 Mpa, and a frequency = 15 Hz. All the samples were analyzed after fracture to determine the exact location of crack initiation from a particular pit. The fracture surface was analyzed to obtain the crack parameters. The stress intensity factor for each sample was calculated using Newman–Raju approximation [19] and crack growth behavior was obtained through CRACK calculation [20].

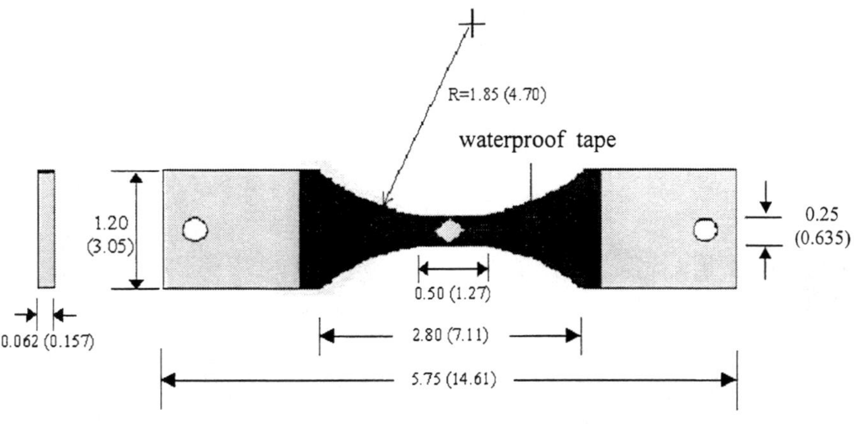


**Figure 1. Geometry of dog-bone fatigue sample made of Al 2024-T3. Primary dimensions are in inches, and dimensions in parentheses are in cm.**

## 3 RESULTS and DISCUSSIONS

All the results discussed are for specimens with L-T orientation. The test matrix was designed to generate a wide range of pit size and depth. Controlled potentiostatic polarization scheme was used to produce a wide range of pit sizes in a relatively short amount of time. Figure 2 shows a representative optical micrograph of a pit in L-T oriented specimen that was anodically polarized with the substrate coated with the imperfect layer of microstop coating. The polarization times range from four hours to twenty-four hours for the six samples shown in Figure 2. For longer polarization times, larger and deeper pits were observed. The pit morphology was characterized using WLIM. Figure 2b to 2d shows the details of pit morphology obtained using different options available with WLIM. Figure 2c shows the maximum contrast while figure 2d provides a detailed 3-D morphology of the pit. SEM was also used to characterize a few select samples. A more extensive display of these profiles can found in reference [20]. Since SEM is a well established method, it was used to mainly compare the qualitative surface features with those shown in the WLIM surface profiles.

The sample shown in Figure 2 was coated with the micro-stop to restrict the number of sites for pit nucleation and to allow formation of a few large pits on the sample surface. A number of specimens were pitted to establish some parameter, which results in one or two dominant pit. These parameters were later used to create controlled pitting, one or some times two, out of which one became dominant with increasing polarization time. The polarization times for the sample ranged from 4 hours to 24 hours. The longer polarization times produced larger pits. As the polarization times increased, the pit depths also increased. However, the sample polarized for twenty hours had deeper pits than the sample polarized for twenty-four hours [21]. In this case, instead of growing deeper, the pits expanded during the extra polarization time. This behavior is shown in Figure 4, which is a plot of polarization time vs. critical pit depth. All data points were determined to be statistically valid using the average deviation method [22].

**Figure 2. Change in Critical Pit Depth with Increasing Polarization Time.**

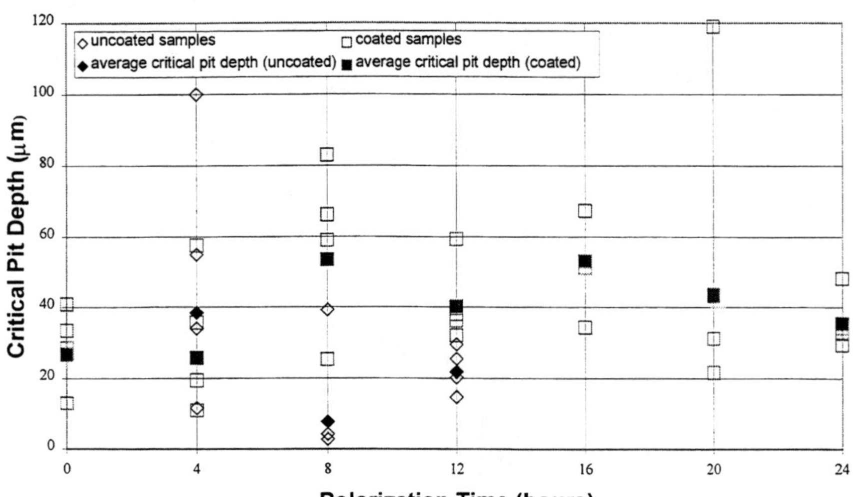

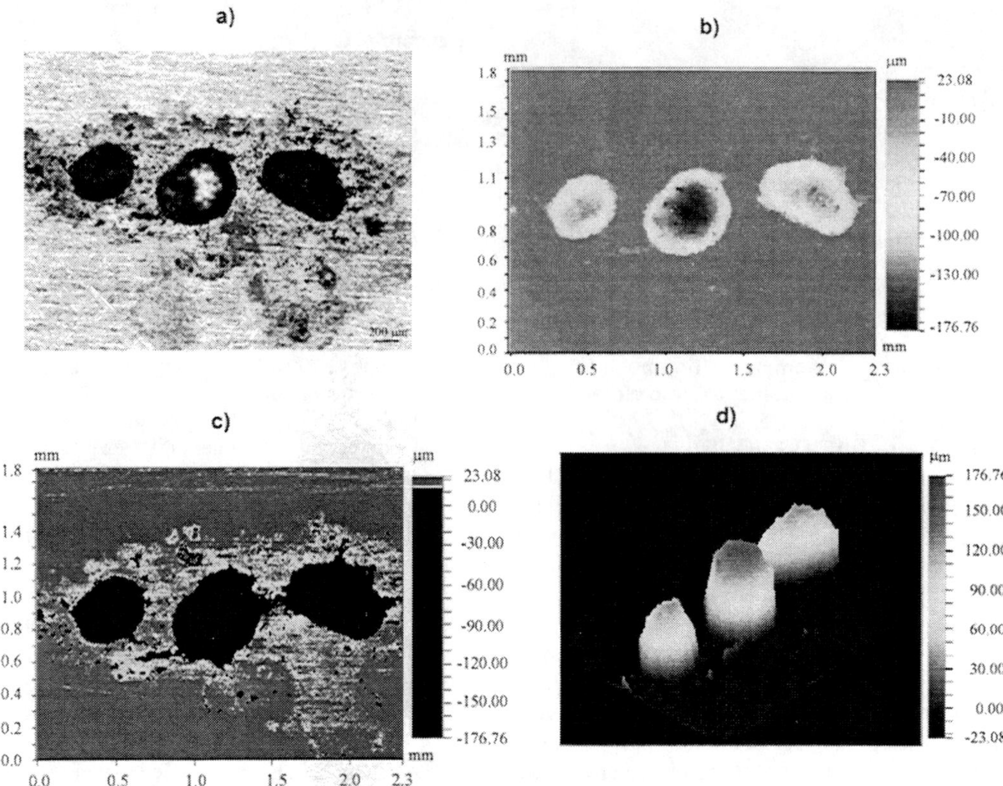


**Figure 3. Characterization of Pit Morphology: a) Optical Micrograph, b) WLIM Surface Profile, c) WLIM Maximum Contrast and d) 3-D Image of the Pits.**

A two-dimensional analysis was performed for each sample using WLIM as described in reference [21] to determine the dimensions of the pits produced on these samples. The pitted samples were fatigued after the surface pit morphology was characterized completely. The samples were fatigue tested at a stress ratio, R = 0.1 and a frequency, f = 15 Hz. The maximum stress in all cases was 256 Mpa. The main goal was to determine a relationship between pit depth and stress intensity factor, K. The critical pit size is assumed as the size of the pit that initiates fatigue crack. Figure 5 schematically illustrates such a fatigue crack initiation and growth process from a pit [23]. This figure represents a flow chart of the experimental procedure. Step 1 simply shows the dog-bone fatigue sample with the load direction and a small area of the gage section where pits are to be initiated. Step 3 shows three pits created by controlled accelerated technique. One of these pits initiates fatigue crack on fatigue loading and a semi-elliptic crack appears to grow. This stage is depicted in Step 3. Step 4 is a two dimensional scheme of the crack initiation and growth process with the two important parameters of the pit depth and diameter (length) used in stress intensity factor calculation.

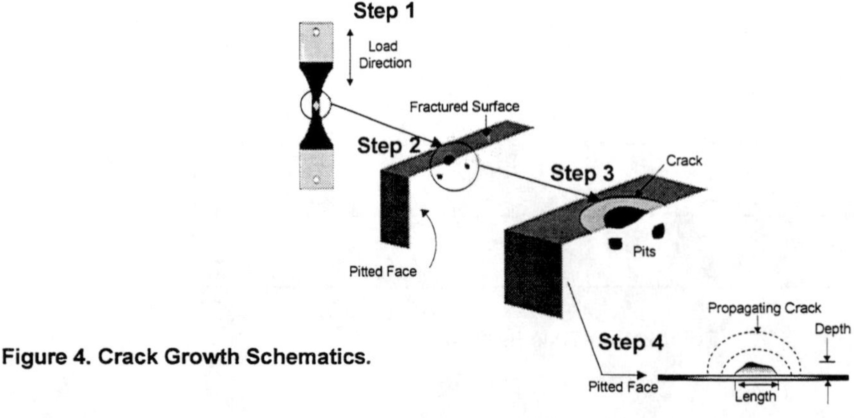


**Figure 4. Crack Growth Schematics.**

A detailed analysis of fracture surface indicated that crack always started from the pit at the corroded surface. In most cases the crack initiated from a single pit and a single semi-elliptical surface crack grew to a critical size leading to the failure of the specimen. In some cases, the cracks initiated from two or more adjacent pits. In these cases, the cracks linked together to form a single semi-elliptic crack that eventually caused the failure. Figures 5 shows several optical micrographs of a sample before and after it was failed under fatigue. The pitted face was examined first to determine the area in which the crack developed. For the sample in Figure 5, it appeared that the crack initiated from both of the pits. The fractured surface was then examined for the initiation site of the crack. The crack appeared to be centered around the left pit. The left pit is the crack initiating pit and it corresponds with the top pit shown in Figure 5(a). The depth and diameter of this pit were calculated using the WLIM two-dimensional analysis.

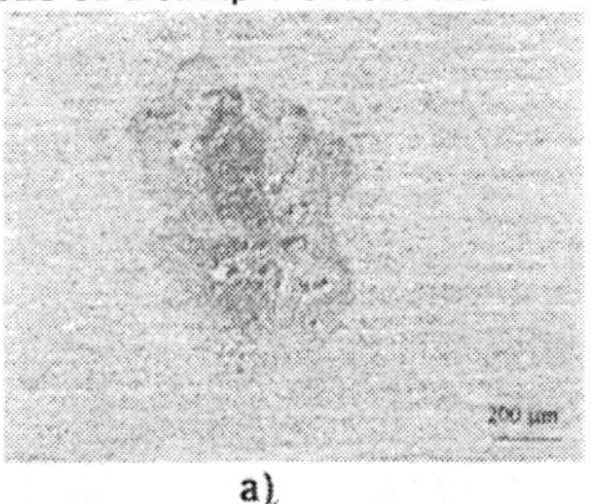

a)

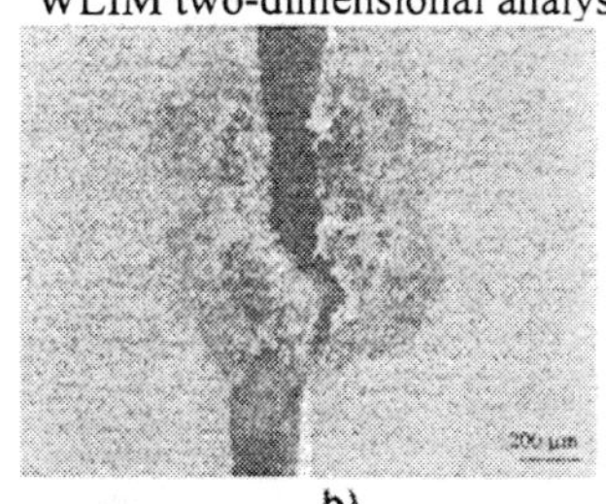

b)

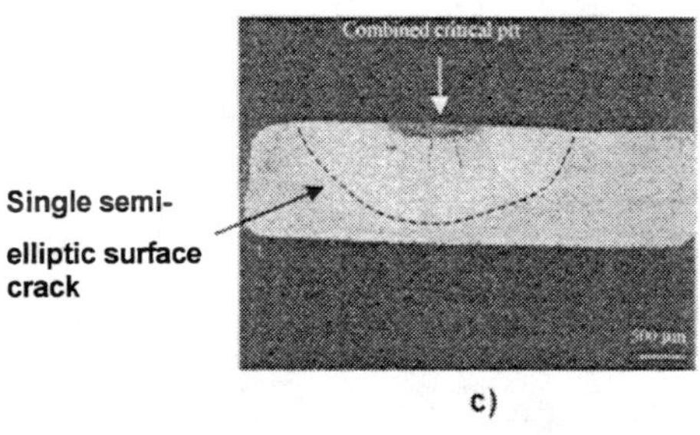


Single semi-
elliptic surface
crack

c)

**Figure 5. Fractured Microstop Coated Dog-Bone Sample Polarized for 20 Hours.**
**a) Optical micrograph of pitted surface before fracture**
**b) Optical micrograph of pitted surface after fracture**
**c) Fractured surface showing semi-elliptic crack**

Once the parameters of interest were obtained using WLIM, the stress intensity factor for each specimen was calculated. The stress intensity factor was used as the key fatigue parameter in developing a relationship between pit depth and fatigue parameter. The stress intensity factor for each sample was calculated using a program developed at the University of Dayton Research Institute [20].

One of the main objectives of this study is to develop a relationship between a structural integrity parameter and a pitting parameter. Such a relationship is shown in Figure 6. The stress intensity factor was plotted against critical pit depth. There appears to be a linear relationship between the two parameters. This relationship can be used to predict the life of structure. Fatigue crack growth model developed by University of Dayton Research Institute CRACKS model was used to predict the crack growth behavior of fatigue cracks initiating from a pit. A single semi-elliptical surface crack is assumed to develop from the critical pit. The model uses the stress intensity factor number from previous calculation and the crack growth data is generated from the initial and final crack size measured. Crack growth is assumed to start from the first applied cycle with no crack initiation period. Such an assumption is a reasonable approach because majority of the fatigue life of pitted sample comprises of growth stage [24].

Crack growth data in the form of da/dN vs. stress intensity factor is shown in Figure 7. The data shown as open squares are taken from notched compact tension specimen representing long crack growth [21]. The closed squares represent the data from pitted samples used in this study. It appears that the crack growth behavior of cracks initiating from pit is typical of short crack growth behavior. However, a transition to long crack growth behavior is quite evident suggesting that pits deeper than 80 μm behave similar to standard notched specimen representing long crack growth data. More tests are required with critical pit depths in the range of 100-200 μm.

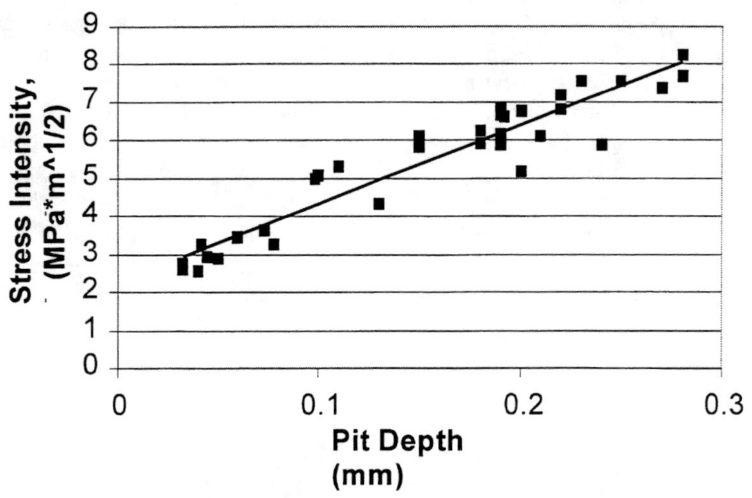


**Figure 6. Stress Intensity Factor Versus Critical Pit Depth.**

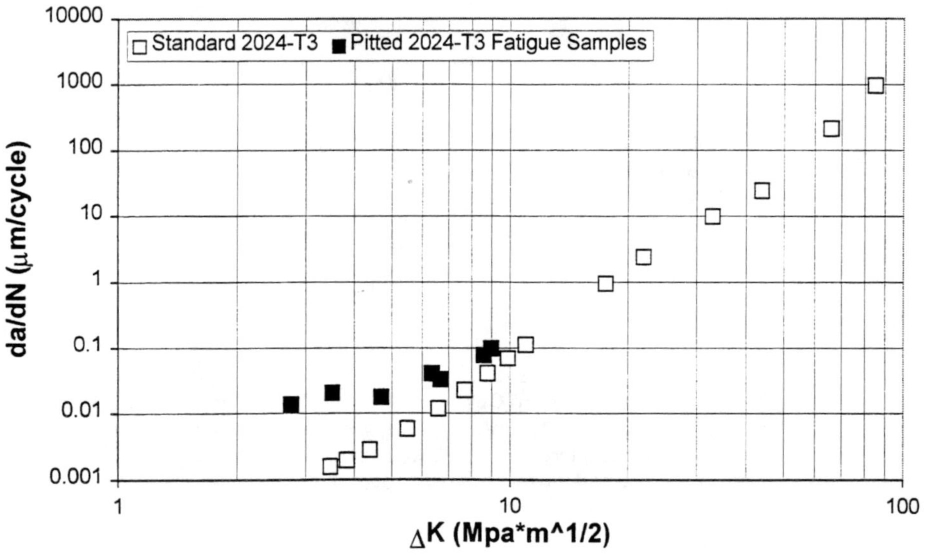


**Figure 7. Crack-Growth Rate Versus ΔK.**

## 4 SUMMARY

A select number of pits were created on dog-bone fatigue sample by an electrochemical technique. The pit depth increased with increasing polarization time. However, this relationship failed at polarization times longer than 20 hours. All the pits were characterized using WLIM and SEM. Important parameters such as length and depth of the pits were determined. The fracture surface was analyzed to determine the initiation and failure locations on the surface. Stress intensity factors were calculated using all the parameters measured. A linear relationship was observed between critical pit depth and K. Fatigue crack growth initiating from a pit was observed to be typical of short crack. Sign of possible transition to long crack growth behavior was observed for pits deeper than 80 μm.

## 5 ACKNOWLEDGMENTS

Effort sponsored by the Defense Advanced Projects Agency (DARPA) Multidisciplinary University Research Initiative (MURI), under Air Force Office of Scientific Research Grant Number F49620-96-1-0442. Technical assistance from Y. Ding, T.E. Matikas, P.C. Mieldar, N. Schell , and L.B. Simon are gratefully acknowledged.

## 6 REFERENCES


1. Rokhlin, S. I., and Kim, J.Y., "Fracture Mechanical Analysis of Fatigue Crack Initiation and Growth from Pitting Corrosion", Second Annual Report for DARPA-MURI Under AFOSR Grant Number F49620-96-1-0442, (1998) p. A24-2.
2. Groner, D. J., "US Air Force Aging Aircraft Corrosion", Current Awareness Bulletin, Structures Division, AFRL/FIBD, Aerospace Structures Information and Analysis Center, WPAFB, OH, Spring 1997.
3. Chen, G.S., Wan, K.C., Gao, M., Wei, R. P., and Flourney. Materials Science & Engineering, A 219, p. 126, 1996
4. Congleton, J., Olieh, R. A., and Parkins, R. N. Met Technol. 9, p. 94, 1982
5. Kondo, Y. " Prediction of Fatigue Crack Initiation Life Based on Pit Growth", *Corrosion*, Vol. 45, no. 1, (1989) p. 7.
6. Ahn, S. H., Lawrence Jr., F. V., and Metzger, N. M. Fatigue Fracture Eng. Mater. Struct. 15, p. 625, 1992.
7. Chen, G. S., and Duquette, D. J., Matall. Trans. 23A, p. 1563, 1992.
8. Huang, Y. H., and Alkire, R. C., J. Electrochem. Soc. 136, p. 2763, 1989.
9. Piascik, R. S., and Willard, S. A. "The Growth of Small Corrosion Fatigue Cracks in Alloy 2024", *Fatigue and Fracture of Engineering Materials and Structures,* Vol. 17, No. 11, (1994) pp. 1247-1248.
10. Harmsworth, Clayton L. "Effect of Corrosion on the Fatigue Behavior of 2024-T4 Aluminum Alloy", ASD Technical Report 61-121, (1961).
11. Perez, R., "Metrics Exploration and Transformation Development for Corrosion/Fatigue", Final Report, Delivery Order 0005, USAF Contract F33615-95-D-3216, November 1996.
12. Mitchell, G., "Application of Corrosion Quantification Techniques to the USAF KC-135 Stratanker", Proceedings of the Air Force 4th Aging Aircraft Conference, United States Air Force Academy, CO, July 1996, p. 251.
13. Wanhill, R. J. H., "Aircraft Corrosion and Fatigue Damage Assessment", Proceedings of the US Air Force ASIP Conference, San Antonio, TX, December 1995, p.9.
14. Cole, G. K., Clark, G., and Sharp, P.K., "The implications of Corrosion with Respect to Aircraft Structural Integrity", Aeronautical and Maritime Research Laboratory, Melbourne, Australia, Research Report DSTO-PR-0102, AR-010-199, March 1997.
15. Nakajima, M., and Tokaji, K., "A Mechanical Condition of Fatigue Crack Initiation from Corrosion Pits", Fatigue 1996, Proceedings of the Sixth International Fatigue Congress, Berlin, Germany, pp. 697-702, 1996.
16. Philips, E. P., and Neuman, J. C., "Impact of Small Crack Effect on Design Life Calculations"" Experimental Mechanics, pp. 221-225, 1989.
17. P. Doig and P.E.J. Flewitt, Mechanisms of Environment Sensitive Cracking of Materials, Metals Society, University of Surrey, p. 113 (1977).
18. T.H. Sanders, Jr., et al., Effect of Microstructure on Fatigue Crack Growth of 7xxx Aluminum Alloys Under Constant Amplitude and Spectrum Loading, Report 56-78-AF8, Contract N00019-76-C-0482, Aluminum Company of America, ALCOA Center, PA, (April, 1978).
19. Newman, J. C., and J. Raju, I. S., "Stress-Intensity Factor Equations for Cracks in Three-Dimensional Finite Bodies", Fracture Mechanics: Fourteenth Symposium-Volume 1: Theory and Analysis. ASM STP 791, (1983) pp. I-238-I-265.
20. John, R., Personal communication, AFRL, Wright Patterson Air Force Base, Ohio.
21. Simon, L. B., "Influence of Pitting Corrosion on the Loss of Structural Integrity in Aluminum Alloy 2024-T3", Master's Thesis
22. Wilson, L., Quantitative Analysis: Gravimetric, Volumetric and Instrumental Analysis, 3rd Edition, Mohican Textbook Publishing Co., Loudonville, Ohio, (1995) pp. 50-58.
23. Shankaran, K. Shankaran., "Validation of Accelerated Corrosion Tests for Aging Aircraft Life Prediction: Effects of Pitting Corrosion on the Mechanical Behavior of 7075-T6 Aluminum Alloy", Final Report, 1998, F33615-96-D-5835, Delivery Order: 0026-02, AFRL, Wright-Patterson AFB, OH.
24. Bray, G.H., Bucci, R.J., Colvin, E.L., and Kulak, M., "Effect of Prior Corrosion on the S/N Fatigue Performance of Aluminum Alloys 2024-T3 and 2524-T3", Effects of the Environment on the Initiation of Crack Growth, ASTM STP 1298, 1997, p. 89.

Proceedings of
2001 ASME International Mechanical Engineering Congress and Exposition
November 11–16, 2001, New York, NY
NDE-Vol. 21

**IMECE2001/NDE-25817**

# USE OF COMPUTATIONAL INTELLIGENCE FOR CORROSION DAMAGE ASSESSMENT OF AGING AIRCRAFT PANELS

**M. J. Palakal**
Department of Computer Science
Indiana University Purdue University
Indianapolis, IN 46202
Email: mpalakal@cs.iupui.edu

**J. Huang**
Department of Computer Science
Indiana University Purdue University
Indianapolis, IN 46202
Email: huang@cs.iupui.edu

**R. M. Pidaparti**
Department of Mechanical Engineering
Indiana University Purdue University
Indianapolis, IN 46202
Email: ramana@engr.iupui.edu

**D. T. Peeler**
Air Force Research Laboratory,
System Support Division, AFRL/MLSA
Wright Patterson AFB, Dayton, Ohio 45385

**S. Rebbapragada**
Department of Computer Science
Indiana University Purdue University
Indianapolis, IN 46202

## ABSTRACT

*This paper describes computational methods that have been developed to automatically detect and quantify corrosion on images of aircraft panels obtained using nondestructive inspection (NDI) techniques. Images obtained from NDI techniques are not directly suitable for corrosion analysis. We propose a multi-resolution wavelet analysis to segment the NDI images to isolate the corroded regions from the undamaged regions. Wavelet coefficients were extracted from each segment and subjected to a multivariate data analysis to obtain the most relevant features. These features are then used to identify the corroded region on the images and also to estimate the extent of damage, in terms of material loss, using backpropagation neural networks. Experiments were conducted to test the developed modules on images obtained from the Air Force Research Lab in Dayton, Ohio. The results obtained prove the feasibility and robustness of our approach in detecting the damaged regions and for estimating the material loss due to corrosion.*

## 1 Introduction

The aging structures phenomenon is one of the most important economic and safety concerns of industrial community. Air transportation is a multi-billion dollar industry world-wide. Aging aircraft structures have various types of damage including corrosion, widespread fatigue damage, and in-service cracking of the aircraft wing upper surface as described by (Lincoln, 1995). Non-destructive evaluation (NDE) procedures involve establishing correlation between measured properties and quantitative information about anomalies. For example, Eddy current NDE involves measurement of impedance and correlating it with the damage (crack, or change in microstructure) producing it. There is a growing demand for improving existing NDE techniques to achieve maximum confidence and reliable results with minimum damage components. Also, due to damage tolerance requirements, it is required to detect and quantify even the smallest crack. Damage due to hidden corrosion and cracks in composites are not always reliably detected using conventional NDE methods. Some of the recent developments in detecting hidden corrosion include tomographic technique (Lanza, 1995), microwave (Martens *et al.*, 1995) and optical technique (Jin and Chiang, 1995), among others. Recently, Lepine (Lepine *et al.*, 1999) developed a pulsed eddy current method for hidden corrosion detection in aircraft structures.

Wallace and Hoeppner (Wallace and Hoeppner, 1985) described many types of corrosion that may occur in aircraft structures. Lap-joints are a common structural element in many military aircraft (KC-135, C-130, Jstars) and in transport aircraft fuselage which is subjected to corrosion damage. Corrosion in lap joints has many effects including material loss, pillowing stresses, pillowing cracks, rivet failures and interactions with fatigue, among others. Corrosion generally attacks the faying surfaces in lap joints, and occurs when the adhesive bond/sealant between the layers breaks down, allowing moisture to penetrate. Pitting corrosion usually follows the general attack of corrosion on faying surfaces and may sometimes evolve into exfoliation corrosion.

The general corrosion attack, pitting corrosion, and exfoliation corrosion are characterized in terms of material (thickness) loss. Fatigue may interact with corrosion in the lap-joints, and its effect is not completely understood. The fatigue life of a corroded faying surface can greatly affect the fatigue life of a joint (Brooks *et al.*, 1999). Pillowing is the deformation of the layers of the lap-joint between the fasteners. Due to pillowing, cracks may initiate from the rivets due to changes in the stress state of the lap-joint, and may control the life of the lap-joint (Bellinger *et al.*, 1998). In addition, pillowing stresses may themselves cause rivet failures (Bellinger *et al.*, 1999). Researchers from the Institute for Aerospace Research (Bellinger, *et al.*, 1999) through NDI techniques found many examples of cracked and failed rivets during disassembly of service-retired aircraft.

Detecting damage using Eddy-current and ultrasound NDE techniques poses a great difficulty because of poor signal to noise ratio of the signals. To improve the probability of detection, advanced signal concepts like spectral analysis and pattern recognition techniques can be used. Currently, these images are mostly used for damage visualization but not for damage characterization and assessment. Corrosion damage assessment using pattern recognition techniques has the potential for automating the NDE process by providing estimates of corrosion damage metrics in terms of morphology, material loss, and improving the structural integrity assessment.

In the past, most NDE for corrosion has concentrated on material loss information, which is not sufficient for structural analysis models, thus limiting the progress of NDE technology. Also, the current "find it - fix it" philosophy has discouraged the use of sensitive NDE techniques by commercial and military operators. Due to rising costs for corrosion related maintenance, alternate corrosion maintenance policies and practices need to be explored to reduce corrosion cost without compromising the safety.

The paper describes computational methods that have been developed to automatically detect and quantify corrosion on images of aircraft panels obtained using nondestructive inspection (NDI) techniques. Since images obtained from NDI techniques are not directly suitable for corrosion analysis, we have developed set of image processing and pattern recognition algorithms to isolate the corroded regions from NDI images and to assess the extent of corrosion damage.

## 2 Structural Corrosion Damage Assessment Methodology

Figure 1 shows the organization of the structural corrosion damage prediction methodology. The development of the process involves interfacing an NDI system with a database, quantification and classification of damage, estimation of the severity of the quantified damage and prediction of the structural integrity in terms of fatigue life, residual strength, and others. The quantified corrosion damage along with the metrics will be used for structural integrity assessment to estimate the life cost. A management decision (repair or replace) can be made based on the corrosion quantification metrics or structural integrity assessment using the impact of repair, replace, or life cost.

As shown in Figure 1, the corrosion damage assessment process consists of four major components: (i) a corrosion damage database module to hold damage images from various NDI systems; (ii) modules for damage identification and quantification using computer vision, pattern recognition, and intelligent computational techniques; (iii) a structural integrity assessment module; (iv) life cost and impact of repair analysis module; and (v) an integrated management decision maker using expert system methodologies to make repair/replace decision. The database will contain corrosion damages indexed with source, quantification, and information about any repair decision made by the management decision-maker. The knowledge base on the other hand will support structural integrity assessment using experimental, probabilistic, and deterministic methods. In this paper we discuss the details about corrosion damage identification and quantification from NDI images.

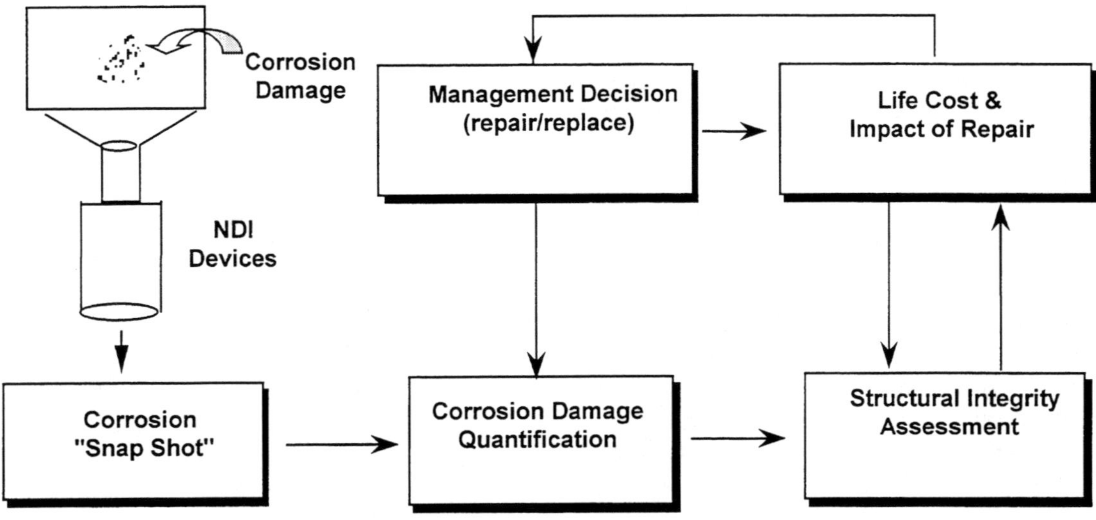


Figure 1. Structural Corrosion Damage Assessment Process Methodology

## 3 Corrosion Damage Identification and Quantification

In this section, we discuss how to quantify the corrosion damage due to different NDI sources in aging structures/components and estimate the severity of the quantified damage using intelligent computational methods.

The overall process of identification and quantification of corroded regions from NDI images involves two stages: first, classifications of various regions in the image as corroded or uncorroded, and second, prediction of the material loss of the corroded regions. The classification process involves segmenting the image into various regions. Multi-resolution wavelet analysis is performed on the NDI images to obtain a set of wavelet coefficients as feature vectors. These features are used for the identification of the damaged regions on the NDI images using clustering techniques. Each of the segments on the segmented image would correspond to a damaged region or an undamaged region. Segmentation of the images of damaged panels is similar to the segmentation of texture images. Though a number of segmentation algorithms are available in the literature, texture segmentation based on wavelets seems to be a robust technique in this particular application. After the image is segmented, histogram and wavelet features are extracted from each segment for damage quantification. The damage identification and quantification steps are described in detail now.

## 4 Corrosion Damage Identification

The images that are obtained through conventional NDI methods are not directly suitable for identification of damaged regions automatically. Therefore, the damage identification process involves three major steps: feature extraction, segmentation, and classification. Wavelet analysis techniques are used for feature extraction, a Clustering technique is used for segmentation, and K-means distance-based method is used for damage classification.

The feature extraction process involves multi-resolution wavelet analysis on the NDI images to obtain a set of wavelet coefficients as feature vectors. This would provide four images, each of which is the size of the original image. Therefore there exists a wavelet coefficient corresponding to every pixel in the original image in all the four transformed images. A feature vector is now formed for each pixel in the original image. This feature vector consists of the pixel gray level intensity in the original image and the corresponding coefficients in the transformed image. In case of color images, each of red, green and the blue channels are considered as three different gray level and the wavelet transforms are performed on each of the three levels. This would hence result in three times the number of features than the gray images. Though the wavelet transforms can be repetitively performed on each of the resulting transformed images, we observed that just two levels of transformation produced a good number of features necessary to achieve a good segmentation.

During the segmentation process, various regions in an NDI image are isolated for classifying these regions into corroded or non-corroded in the next step. Image segmentation is an important step in this whole process. Traditionally, segmentation algorithms are based on one of the two basic properties of gray-level values: discontinuity and similarity. These methods are usually classified as thresholding, edge detection or gradient based, region growing, and hierarchical scheme (Unser and Eden, 1989). In our approach, the wavelet features obtained from earlier steps are used for segmentation. Segmentation is achieved by classifying two neighboring pixels into the same segment when they belong to the same cluster. A K-Means Clustering Algorithm (Rebbapragada, 2000) is used for segmenting the images.

Once the image is segmented to a certain number of levels, the next task is to distinguish the segments that represent the corroded regions from those which represent the non-corroded regions. A wide variety of features can be extracted from each of the segments and can used to identify the damaged regions. Binary object features such as area of each segment, perimeter, thinness ratio, aspect ratio can be calculated. The histogram features can include the mean of the pixel values in the segment, standard deviation, skew, energy, and entropy (Umbaugh, 1998). These features form the feature-vector for classifying the corroded and the non-corroded regions. Since the number of segments that were formed is usually very high, only mean and standard deviation using the lowpass/lowpass wavelet coefficients was used.

## 5 Corrosion Damage Quantification

Once the damaged regions are identified on the image, the next task is to quantify these damages. The extent of material loss is used to quantify the damages on the panels. Various features that are obtained on the corroded region based on histogram and/or wavelet analysis during the segmentation stage will be used to estimate the material loss.

First order statistical features will be computed using the histogram of the NDI images. These first order features include (i) mean, (ii) standard deviation, (iii) skew, (iv) energy, and (v) entropy. The first order histogram probability $P(g)$ is defined as $P(g) = N(g)/M$, where $M$ is the number of pixels in the image and $N(g)$ is the number of pixels at gray level[26] $g$. Then, the first order features can be calculated as,

(i) Mean, the average value is defined as,

$$\bar{g} = \sum_{g=0}^{L-1} gP(g) = \sum_r \sum_c \frac{I(r,c)}{M}$$

(ii) Standard Deviation defined as,

$$\sigma_g = \sqrt{\sum_{g=0}^{L-1} (g-\bar{g})^2 P(g)}$$

(iii) Skew measuring the asymmetry as,

$$\frac{1}{\sigma_g^3} \sum_{g=0}^{L-1} (g-\bar{g})^3 P(g)$$

(iv) $$\text{Energy} = \sum_{g=0}^{L-1} [P(g)]^2$$

(v) $$\text{Entropy} = -\sum_{g=0}^{L-1} P(g)\log_2[P(g)]$$

The second order features are calculated using a co-occurrence matrix. The co-occurrence matrix is an estimate of the second order joint probability density. The joint probability is $P$(i,j,d,t), the probability of grey levels $i$ and $j$ occurring at distance $d$ and angle $t$. A co-occurrence matrix for distance $d$ and angle $t$ is composed of elements $g_{ij}$ which record the number of pixel pairs $d$ units apart in the direction $t$, with one grey level at $i$ and one at $j$, the order being immaterial. The co-occurrence matrix can be used to calculate the second order features as,

Angular Second Moment,

$$\text{ASM} = \sum_i \sum_j [g(i,j)]^2$$

Inverse Difference Moment,

$$\text{IDM} = \sum_i \sum_j \frac{1}{(j-i)^2} g(i,j)$$

$$\text{Entropy} = \sum_i \sum_j g(i,j)\log(g(i,j)$$

$$\text{Contrast} = \sum_i \sum_j (i-j)^2 g(i,j)$$

An artificial neural network (ANN) is now used for the prediction of material loss using the first and second order features. ANNs are capable of realizing a variety of non-linear relationships of considerable complexity. The ANN used for predicting material loss has input nodes equal to the number of features extracted from each segment. The number of input nodes used is based on the number of features considered. There will be only one output node for the network indicating the material loss in the segment to which the feature-vector corresponds. The set of weights is initially assigned randomly. The training of the network involves adjusting the weights so to reduce the error within some predefined limit. After a good number of training examples, the network would then be used to predict the material loss due to corrosion.

## 6 Results and Discussion

A number of different segmentation methods such as the PCT/Median, SCT/Center, and wavelet-based algorithms were performed on the corrosion damage image data. It was observed that the segmentation performed based on wavelet parameters is often superior to the segmentation performed using the standard segmentation algorithms for NDI image. Another advantage of using wavelets is that the feature vectors formed for each pixel can be used for further analyzes of each segment. Figure 2 shows a sample segmentation result obtained from a corroded image using Eddy current. The gray regions between the rivet holes are the corrosion that is clearly identified through the segmentation process.

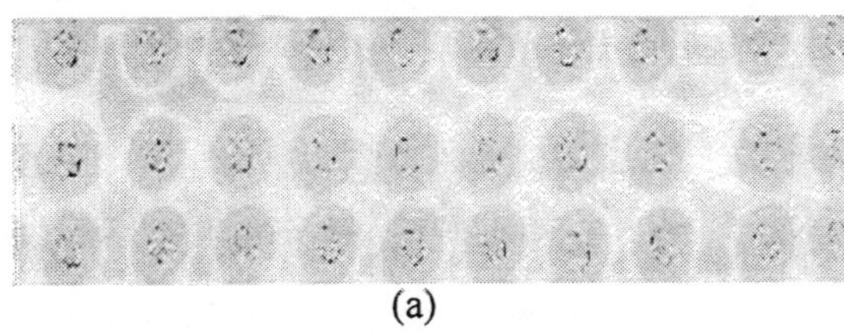

(a)

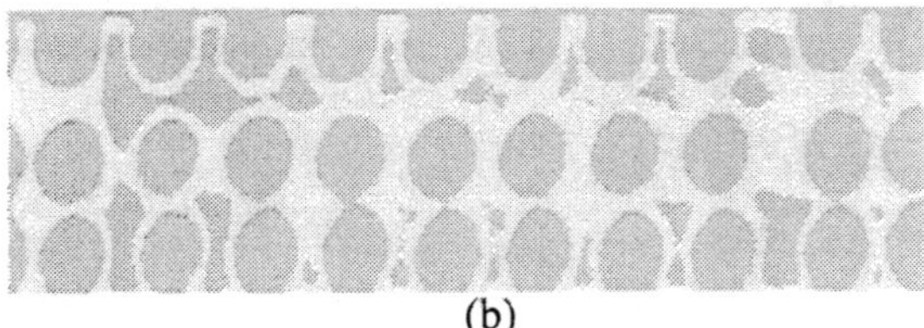

(b)

Figure 2. Segmentation of an eddy current image performed using wavelet parameters. (a) the original image. (b) segmented image using wavelets.

Figure 3 shows the identification and quantification capabilities of the system on eddy current images of another panel. The original image is obtained at different frequencies. Calibrated specimens were used to train the neural network to identify and quantify damages.

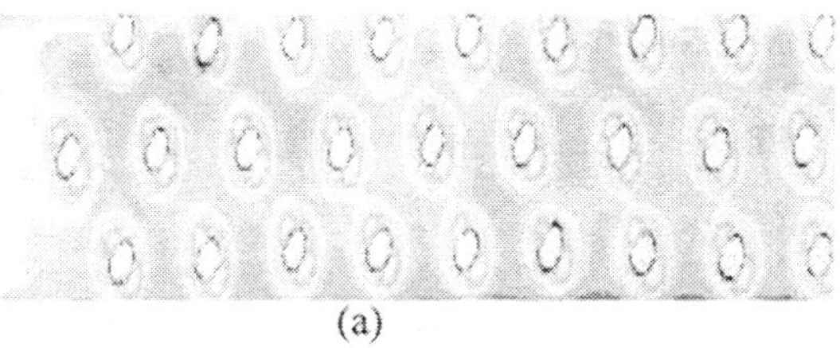

(a)

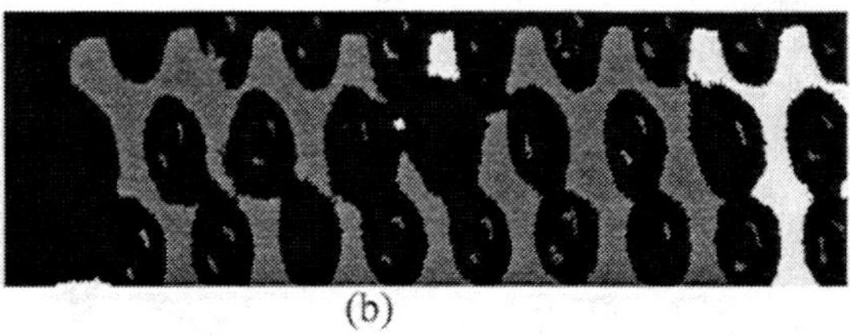

(b)

Figure 3. Preliminary results on corrosion data obtained from the National Research Council of Canada (courtesy of Dr. Peeler of AFRL and Dr. Forsyth of NRC). The original damage panels (a) were obtained using Eddy Current at 12kHz. Image (b) is the identified and quantified corroded regions obtained using the proposed techniques. In this image, the dark region represents 0% corrosion where as the lightest region corresponding up to 25% corrosion.

Quantized information about the damage can be used for further analysis such as severity evaluation, prediction, repair guidance, and so on.

## 7 Concluding Remarks

This paper proposes a technique for automatic corrosion identification and quantification. The methodology gives an indication of the extent of damages, in terms of material loss, based on the images that are obtained using NDI techniques. Since visual inspection of structure or the NDI images is a tedious and an unreliable task, this method would be of immense help to the material inspection personnel and would also raise safety standards.

The corrosion assessment algorithm methodology is a product that brings together the developments made in image processing and signal analysis, with artificial intelligence capabilities, that would prove invaluable to safety inspection personnel. An appropriate image enhancement algorithm was chosen to enhance the images since they are not suitable for direct processing. Wavelets analysis, which has the capability to reveal aspects of data such as trends, breakpoints, discontinuities and similarity, were used both in segmentation of NDI images and also in extracting features that are superior to the histogram features. Neural networks are capable of realizing a variety of non-linear relationships and are hence exploited to provide artificial intelligence capabilities to the corrosion estimation process.

The long-term goal of the proposed research is also to further extend the developed methods augmented with damage quantification techniques to further develop an intelligent structural integrity prediction system for long-term durability studies of aging structures and materials in other areas such as in transportation systems, bridges, and heavy machinery.

**Acknowledgement**

The authors thank the National Science Foundation for supporting this work through a grant CMS-9812723 with Dr. Ken Chong as the Program Director.

**REFERENCES**

Bellinger N.C., Komorowsski, J. P. and Gould, R.W. "Damage Tolerance Implications of Corrosion Pillowing on Fuselage Lap Joints", Journal of Aircraft, 35, No. 3, pp 487-491, May-June 1998.

Bellinger, N.C., Gould, R.W. and Komorowski, J.P. "Repair Issues for Corroded Fuselage Lap Joints", To be Published in the proceedings of the Third Joint FAA/DoD/NASA Conference on Aging Aircraft, Albuquerque, New Mexico, September 20-23, 1999.

Bellinger, N.C., Gould, R.W. and Komorowski, J.P. "Repair Issues of Corroded Lap Joints" 1999 World Aviation Conference., San Francisco, CA, October 19-21, 1999.

Brooks, C.L., Prost-Domasky, S. and Honeycutt, K. "Corrosion is a Structural and Economic Problem: Transforming Metrics to a Life Prediction Method", Fatigue in the Presence of Corrosion, Report AC/323(AVT) TP/8; RTO-MP-18, North Atlantic Treaty Organization, Brussels(Belgium), Mar 1999

Jain, A.K. and Dubuisson, M.P., "Segmentation of X-ray and C-scan images of fiber reinforced composite materials", *Pattern Recognition*, 25 pp.257-270. 1992.

Jin, F. and Chiang, F.P. "NDE of crevice corrosion using electronic speckle pattern interferometry and digital speckle correlation", ASME AD-Vol. 47, Structural Integrity in Aging Aircraft, pp. 221-235, 1995.

Laine, A. and Fan, J., "An Adaptive Approach for Texture Segmentation by Multi-channel Wavelet Frames", Center for Computer Vision and Visualization, 1993.

Lanza, R. C. "The use of neutraon tomographic techniques for the detection of corrosion damage in aircraft structures", ASME AD-Vol. 47, Structural Integrity in Aging Aircraft, pp. 253-259, 1995.

Lepine, B.A., Wallace, B.P., Forsyth, D.S. and Wyglinsky, A. "Pulsed Eddy Current Method developments for Hidden Corrosion Detection in Aircraft Structures," CSNDT Journal, Vol. 20, No. 6, pp. 6-15, November 1999.

Lincoln, J.W. "The USAF approach to attaining structural integrity of aging aircraft", ASME AD-Vol. 47, Structural Integrity in Aging Aircraft, pp. 9-19, 1995.

Martens, J.S., Sachtjen, S. and Sorensen, N.R. "NDI using MM-wave resonant techniques", ASME AD-Vol. 47, Structural Integrity in Aging Aircraft, pp. 265-271, 1995.

Rebbapragada, S."Intelligent Computational Methods for Corrosion Damage Assessment", Masters Thesis, Indiana University Purdue University, Indianapolis, Indiana, 2000.

Smith, F.G., Jepsen, K.R. and Lichtenwalner, P.F., "Comparision of neural network and Markov random field image segmentation techniques", in Thompson, D.O. and Chimenti, D.E. (ed.) *Review of Progress in Quantitative Nondestructive Evaluation*, Vol. 11A, pp.717-724, Plenum Press, 1992.

Umbaugh, S., "Computer Vision and Image Processing", *A practical approach using CVIP tools*, Prentice Hall, 1998.

Unser, M. and Eden, M., "Multiresolution Feature Extraction and Selection for Texture Segmentation", *IEEE Transactions on Pattern Analysis and Machine Intelligence*, Vol. 11, No. 7, pp. 717-728, 1989.

Unser, M., "Texture Classification and Segmentation using Wavelet Frames", *IEEE Transactions on Image Processing*, Vol. 4, No. 11, pp. 1549-1560, 1995.

Wallace, W., Hoeppner, D. W. "AGARD Corrosion Handbook Volume I Aircraft Corrosion: Causes and Case Histories", AGARD-AG-278 Volume 1, 1985.

**Proceedings of**
**2001 ASME International Mechanical Engineering Congress and Exposition**
**November 11–16, 2001, New York, NY**
**NDE-Vol. 21**

# IMECE2001/NDE-25818

# STUDY OF CORROSION DAMAGE UNDER PROTECTIVE COATINGS

**Mohammad Khobaib[1],**

**Jochen Hoffmann, Shamachary Sathish[1], and Michael S. Donley[1]**

[1]Air Force Research Laboratories, Wright Patterson Air Force Base, Ohio
University of Dayton Research Institute, Dayton, Ohio

## ABSTRACT

Polymer coatings provide an excellent corrosion barrier for Al-skinned military aircraft. However, the degradation and damage of the coatings in their service life over time leads to the initiation of corrosion damage at the substrate level. Early detection and negation of such activity can provide extensive cost savings. Several Electrochemical techniques and Non Destructive Evaluation (NDE) show promise in detecting the onset of corrosion under such coatings. Current accelerated testing of aircraft coating systems for corrosion protection relies heavily on salt spray methods. Electrochemical techniques such as Electrochemical Impedance Spectroscopy (EIS) and Electrochemical Noise Methods (ENM) provide insight into the global properties of a coating system, and both techniques are being used on a limited basis. However, there is a need to investigate corrosion events with greater spatial resolution under coatings at the metal/coating interface. Such corrosion activity may be related to coating defects and variations in the surface chemistry of the underlying metal. The Scanning Vibrating Electrode Technique (SVET) has been developed to allow the investigation of localized corrosion activity with high spatial resolution. Such activity may be associated with coating defects or galvanic coupled regions of the metal surface.

Electrochemical and NDE techniques were used to investigate the early stage of corrosion activity under protective coatings. Coatings in this investigation ranged from a simple epoxy amine to commercially used military aircraft polyurethane coatings. SVET testing of panels with intact high-resistance barrier coatings could not reveal corrosion damage under normal testing conditions because of little or no corrosion activity within the limited exposure time. Chemical, mechanical, and electrochemical means of accelerating the corrosion damage were utilized to obtain results in a reasonable time frame. Corrosion initiation and its progress under the coating were studied in detail and the results are discussed here. Complimentary high-resolution NDE techniques, such as Scanning Acoustic Microscopy (SAM) and Fan Thermography measurements were used to identify the corrosion sites. The overall objective of this investigation is to establish a correlation between the electrochemical and NDE techniques.

## 1 INTRODUCTION

Polymer coatings provide an excellent corrosion barrier for Al-skinned military aircraft. These coatings constitute multi-layer with different additives to satisfy all the requirements of military aircraft coatings. Current coating-systems consist of surface pretreatment (chromate conversion coating), primer (mainly epoxies pigmented with chromates) and topcoat (polyurethanes) [1-3]. The corrosion protection is mainly provided by the surface pretreatment and primer through a complex mechanism of inhibition and barrier functionality. The surface treatment/primer coatings are intended to remain intact throughout the entire Programmed Depot Maintenance Cycle. However, the degradation and damage of the coatings in their service life over time leads to the initiation of corrosion damage at the substrate level. The loss of such integrity results in the corrosion initiation, mainly due to loss of adhesion. Other factors contributing to corrosion initiation include defects, pores in the primer coating, and polymer degradation and resulting changes in transport properties of primer coatings, which allows the ingress of environment to substrate, and, of course, a corrosion prone multi-component metallic surface. All these factors can be in a broader sense classified under loss of adhesion or open substrate at the primer/surface pretreatment interface. Such ingress of environment to the substrate is a key element in the corrosion initiation process. Early detection and negation of such activity can provide extensive cost savings. Nondestructive characterization of corrosion protective coatings is of exceptional importance to enhance usability of coating systems and thus, to reduce the costs of aircraft maintenance and to avoid extensive environmental pollution. The possible failure of the coating leads to insufficient corrosion protection and finally to the failure of the Al alloy. Determination of mechanical and physical properties of the coating and the condition of the interface are required, both under different types of load and environmental exposure. Furthermore, it is necessary to detect corrosion damage in the aluminum alloy below the coating. To achieve this, the employment of state of the art electrochemical technique and high-resolution nondestructive evaluation (NDE) is necessary

Several Electro chemical sensors and Non Destructive Evaluation (NDE) techniques show promise in detecting the onset of corrosion under such coatings. Current accelerated testing of aircraft coating systems for corrosion protection relies heavily on salt spray methods. Electrochemical techniques such as Electrochemical Impedance Spectroscopy (EIS) and Electrochemical Noise Methods (ENM) provide insight into the global properties of a coating system, and both techniques are being used on a limited basis [4-8]. However, there is a need to investigate corrosion events with greater spatial resolution under coatings at the metal/coating interface. Such corrosion activity may be related to coating defects and variations in the surface chemistry of the underlying metal. The Scanning Vibrating Electrode Technique (SVET) has been developed to allow the investigation of localized corrosion activity with high spatial resolution [9-12]. Such activity may be associated with coating defects or galvanic coupled regions of the metal surface.

The SVET has been developed to allow in-situ examination of localized corrosion activity through the detection of minute D.C. current variations associated with corrosion. The technique offers high resolution in current measurements of the order of 0.5 $\mu A/cm^2$ and is able to detect the initiation and progression of corrosion activity under a protective coating. A video image of the sample, captured after each scan, pinpoints the site of corrosion initiation, and calculated current density vectors can be superimposed on the video image to determine the precise location of a corrosion current source. A repeat scan function quantifies corrosion progression with time by allowing the user to choose multiple consecutive scans and observe the growth or repassivation of corrosion sites. The difference in initial corrosion activity under various coatings can be correlated to the failure life of the coatings.

High-resolution NDE techniques, such as SAM [13,14] and Fan Thermography [15] offer opportunities to characterize corrosion events on high strength aluminum alloy airframe structures under polymeric protective coatings. SAM can be utilized to map either coating or interface properties (C-scans). A detailed analysis of such scans provides information on the curing quality of the coatings. It is also possible to detect small corrosion pits under delaminated areas. Furthermore, the reflections of surface waves, which are generated and detected by the same probe enables the examination of the substrate/coating interface which can lead to the detection of sites of weak adhesion or corrosion pitting underneath an intact coating. Flash thermography is a well-established method to characterize defects under coatings. Fan Thermography (hot air heating) uses all the principles of flash thermography with additional advantage of its simplicity and lower cost. The flash lamp is replaced with hot air gun or fan. The IR camera observes the heating process of the material. In comparison to the flash lamps this heating method deposits considerably less energy per time on the surface of examination. The sample heats up much slower, which allows observing the material response over longer periods of time. The lower heat deposition energy reduces the risk of destroying the coating.

In this study, early indications of corrosion under a protective coating was detected and mapped by SVET. Some of these corrosion sites were allowed to grow at different levels of corrosion. SAM and Fan Thermography to detect the loss of adhesion and the presence of corrosion under the coating then characterized these samples. The findings were complimentary and at the same time the combined data provided extensive details about the state of the substrate under the coating.

## 2 EXPERIMENTAL DETAILS

The state of corrosion under a coating was studied initially by mapping local current density of several proprietaries and a few known laboratories formulated coatings through the use of SVET. The coatings were exposed to Harrison's solution, an electrolyte containing 5.00 grams NaCl and 35.00 grams $(NH_4)_2SO_4$ in 1.00 liter of water. In each instance, Al 2024-T3 served as the substrate. Initially, the specimen coatings were intact (with no artificial damage), and the progress of corrosion activity was recorded by scanning a small area within the cell. The cell was formed by gluing a 1/8" high, 2" diameter slice of acrylic pipe to the coated sample surface.

It was noted that in case of high barrier coating systems, the coating performed so well that, under normal immersion, no corrosion damage was detected, even after weeks of continuous electrolyte exposure. In order to establish an accelerated test methodology and obtain results in a reasonable amount of time, corrosion damage would have to be intentionally initiated, allowing the subsequent observation of corrosion progression. As a result, various chemical, mechanical, and electrochemical methods were used to artificially damage the coating and/or the metal substrate to study corrosion initiation. These techniques included placing drops of 10 % $CuCl_2$ under the coating, scribing the sample, and applying an external potential. These studies were carried out with Al 2024-T3 panels, coated with a clear, water-based epoxy primer. The coated samples were characterized by ASM and Fan Thermography after a complete SVET scans and analysis of the specimens which were given various degrees of corrosion.

## 3 RESULTS & DISCUSSIONS

SVET has the advantage of locating the site of corrosion, because of its ability to produce one to one correspondence with the corrosion current measurement and the active sites in the coating. This is achieved by obtaining a video image of the sample captured immediately following a SVET scan and placing it in 1:1 correspondence next to a 2-D current vector plot.

The details of this are described in a previous paper [16] where the two different regions – an intact and a delaminated area in a degraded coating was clearly established by SVET.

Figure 1 shows a 2-D current density mapping of a coated aluminum panel with currently used coating for U.S. Air Force aircraft. The specimen was exposed for over two weeks with maximum current density level of less than 25 $\mu A/cm^2$. This was repeated with three samples and the results were very similar. It quickly became evident that under normal testing conditions, little corrosion damage was detected in panels coated with a high resistance barrier coating, even after prolonged electrolyte exposure.

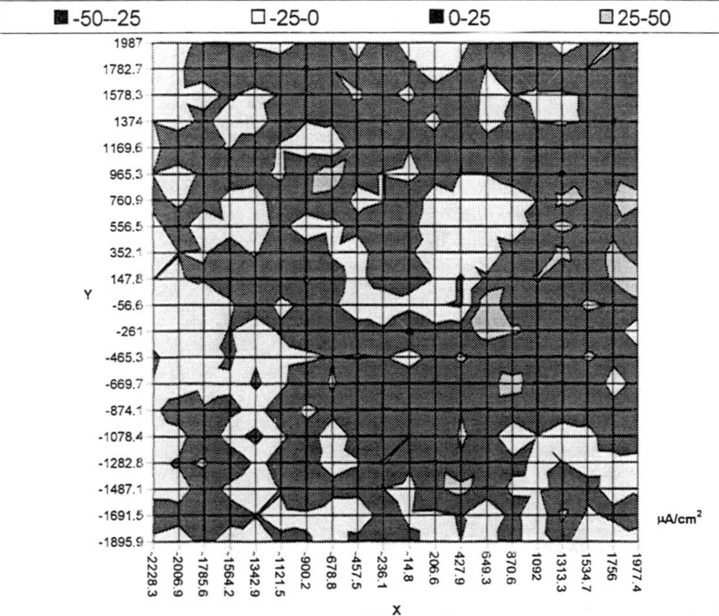


**Figure 1. 2-D Current Density Map of Corrosion Activity under a High Barrier Resistance Coating.**

To obtain reasonable results in a shorter period of time, the creation of artificial corrosion damage was planned. Placing a small drop of an aggressive salt solution on the bare alloy surface was investigated as one method of chemically creating damage under the coating. A drop of 0.2% $CuCl_2$ was placed on the metal panel, which was then coated with a clear, water-soluble primer (thickness = 26-29 μm). Although the drop of salt solution was not visible under the clear coating, each sample was masked and positioned on the SVET sample holder such that the location of the salt drop was in the approximate center of the scanned region.

Figure 2(a) shows the 2-D current density mapping of a sample after 7 hours of total electrolyte exposure. The cathodic region in the center corresponding to the salt drop increased in size with exposure time. In addition, the magnitude of both the anodic and the cathodic current densities increased. Figure 2(b) shows the corresponding vector overlay diagram. In the video image, which was captured immediately after the scan, the coated panel with the 0.2% $CuCl_2$ drop was visible. The dark circular region in the approximate center of the scanned region represents the drop. Comparison of the 2-D mapping after 7 hours of exposure with the corresponding vector overlay clearly identifies the sites of corrosion. However, the corrosion activity did not proceed after exposure of several more days. Hence, a scribe was introduced into the center of the panel as a mechanical means of creating artificial corrosion acceleration.

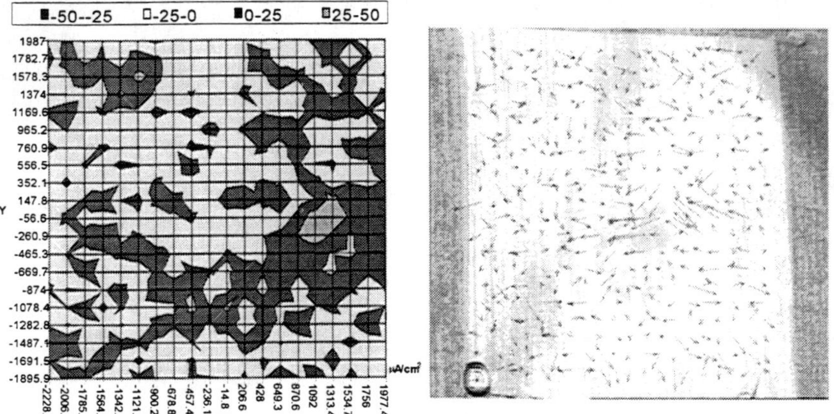


**Figure 2. Corrosion Current Map Under a Proprietary Coating a) 2-D Current Density Map and b) Current Vector Over-lay Imposed over Video Image of the Scanned Area.**

Figure 3(a) shows a scan of the previously described sample taken 2 hours after $CuCl_2$ addition (112 hours total electrolyte exposure). The 2-D current density mapping revealed that the magnitude of both the anodic and cathodic current densities had further increased. In addition, the area corresponding to cathodic current density had also increased in size. Figure 3(b) shows the corresponding vector overlay that was created using the SVET software. In the video image, which was captured immediately after the scan, the coated panel containing the $CuCl_2$-filled scribe is visible. The length of each vector is determined by the magnitude of the current density. The vector tail represents the origin of the current source. Vectors that point upward represent anodic current density while vectors that point downward represent cathodic current density. Comparison of the 2-D scan after 112 hours with the corresponding vector overlay verified that the Cu-containing region was cathodic while the adjacent area was anodic. This analysis provided a complete detail of the corrosion activity under a coating although no visible corrosion was present.

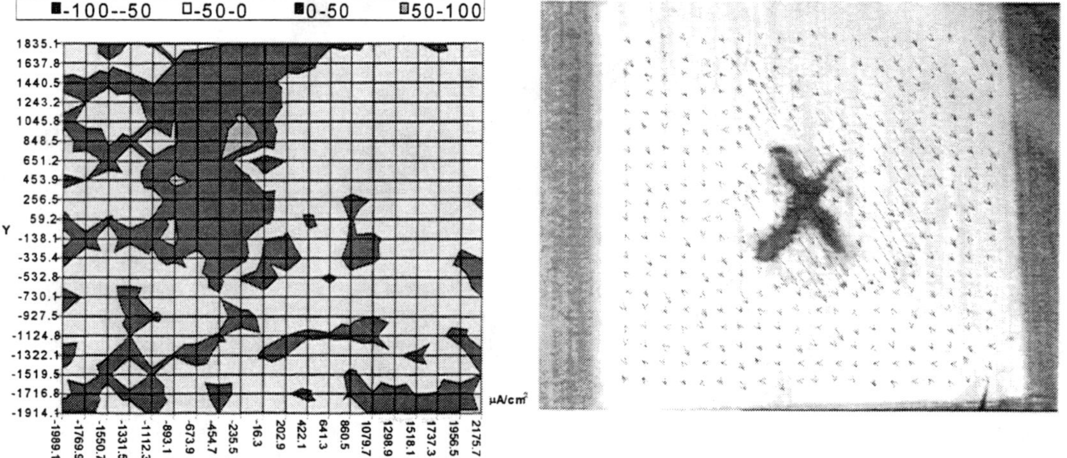


**Figure 3. Current Density Map around Scribed Region: a) 2-D Current Density Map and b) Current Vector Over-lay Imposed over Video Image of the Scanned Area.**

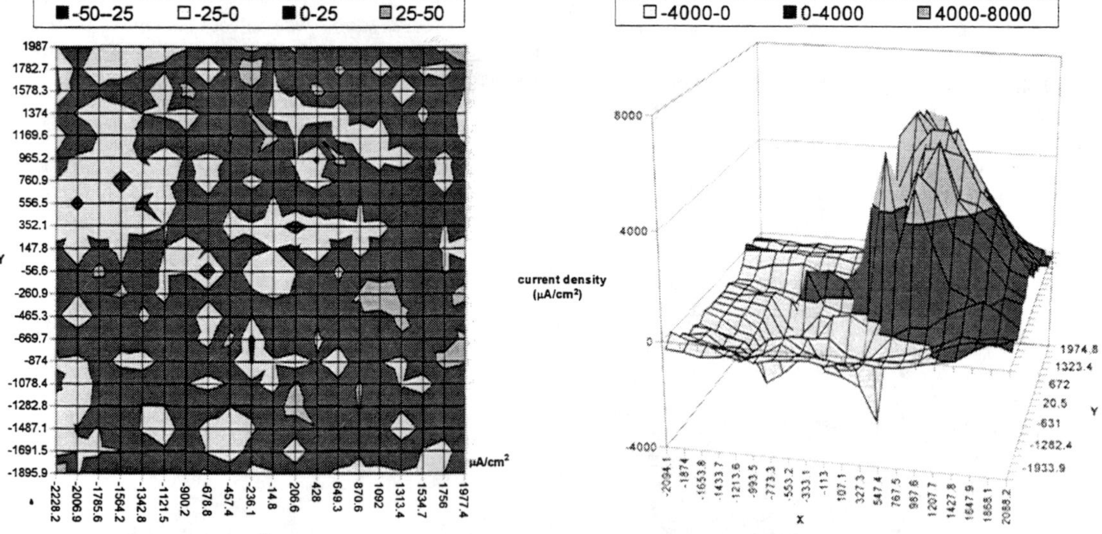


**Figure 4. Current Density Maps Under a Proprietary Coating: a) 2-D Current Density Map at Open Circuit Potential and b) 3-D Current Density Map Obtained After Imposing External Potential.**

Again, in another attempt to study corrosion activity under a coating, a sample with currently used coating for U.S. Air Force aircraft was exposed to Harrison's solution. The current density mappings obtained between 7 and 25 hours of total electrolyte exposure revealed no significant changes in current activity as shown in Figure 4a, so an external potential was applied to electrochemically accelerate the corrosion activity at the substrate. Figure 4b shows the resulting 3-D current density mapping of the sample after applying a potential at 25 hours of total electrolyte exposure. The inverted cone-like shape corresponds to the salt drop region. The magnitude of the anodic current density increased significantly to almost 8000 μA/cm$^2$. Upon applying the potential (polarizing in anodic direction), the coating "broke" and violent bubbling occurred at the drop site. Again SVET provided complete details of the corrosion activity even under very accelerated conditionIt has to be emphasized that the corrosion activity under coating with short exposure is possible even without artificial acceleration. However, this requires coatings with conductive pathways and defects [16]. . Figure 5a and 5b shows the progress of corrosion under such coating with exposure time. The current activity (anodic and cathodic) is spread over the whole panel. The level of the current density at the most active site is over 700 μA/cm$^2$ as shown in Figure 5a, which increased to nearly 1250 μA/cm$^2$ after 18 hours of exposure, as shown in Figure 5b. These data suggest an increase in current activity with extended exposure to Harrison's solution. The corrosion current continued to increase with longer exposure time, indicating the presence of conductive pathways in the coating. This trend was observed with most of the coatings, which had some defects. In the interest of time, such coatings were used to continue NDE investigation.

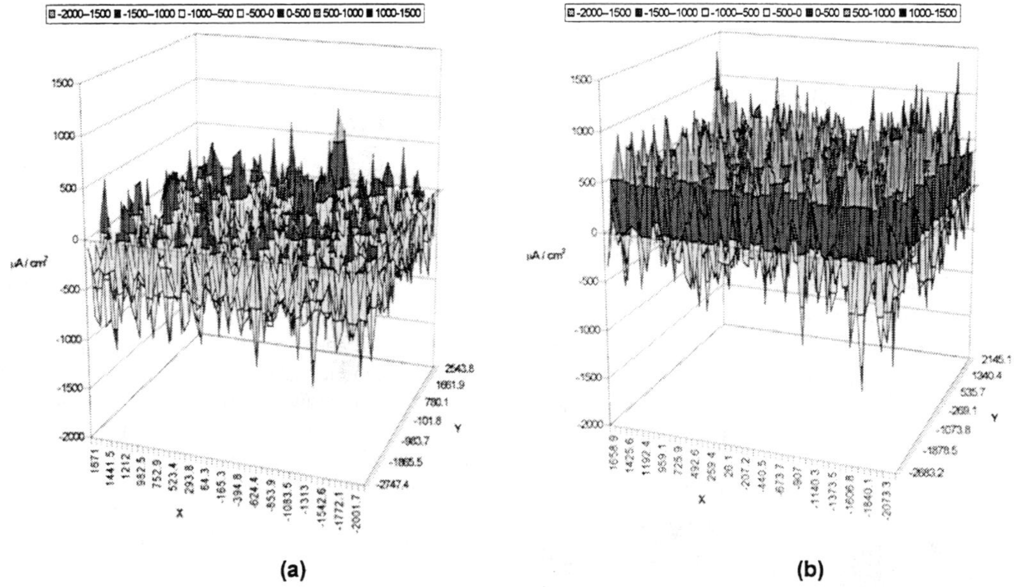


**Figure 5. 3-D Current Density Maps Showing Progress of Corrosion Activity with Exposure: a) 5 hours and b) 18 hours in Harrison's Solution.**

Since corrosion pits can be detected with SAM, initially a correlation of electrochemical mapping obtained by SVET was attempted with the acoustic C-scans. SAM was employed to obtain C-scans on samples with different thickness or interface defects. The possibility of gating certain signals enabled us to measure amplitude variations over the scanning area for different reflections. Thus, several C-scans were obtained per measurement improving the correspondence between the measurement and the material property of interest. Representative C-scans and corresponding electrochemical scans for a 38 μm thick proprietary epoxy coatings (90 hours exposure to Harrison's solution) are presented in Figures 6 and 7. The difference between the two samples is the application of a newly developed thin sol-gel interface layer prior to coating the epoxy primer. The sol-gel layer includes pre-formed, self-assembled nano-phase silane particles (SNAP). These SNAP coatings are considered to enhance corrosion resistance and adhesion to the primer [17]. Results indicate that the SNAP-coatings basically improve adhesion. Figure 6a, shows several spots indicative of delamination and hidden corrosion [18]. The delamination sites were later observed to change to corrosion sites. However, the measured current density in SVET scan does not clearly show distinguishable current density at these sites. The current density however is noted to be higher at the upper two corners. The low level of current density spread over most of the region is indicative of some corrosion activity, but was not distinguishable either by SAM or SVET. However, it is an indication of the poor quality of the coating.

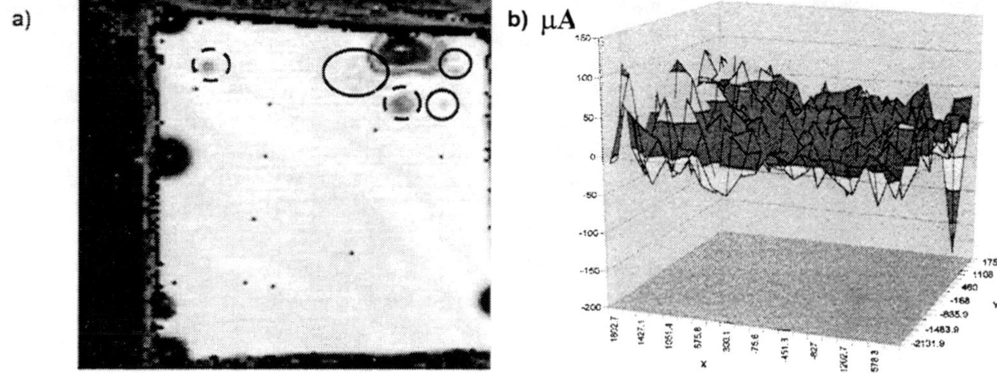


**Figure 6. 38 µm thick epoxy coating, <u>no SNAP layer</u>**
**a) SAM C-scan scan area: 6x6 mm$^2$, stepwidth: 50 µm, substrate reflection**
**Full line circles: hidden corrosion - - dashed circles: hidden delamination**
**Large visible delamination in upper right corner**
**b) SVET scan: 5x5 mm$^2$, 20x20 steps**

In comparison, the SVET scan for the sample with additional SNAP layer shows no significant current density over almost the complete scan-area, except one cathodic (negative) peak (approx. –500 µA/cm$^2$) and one anodic (positive) peak (approx. 400 µA/cm$^2$). The SAM, C-scan clearly detects these spots, but at the same time reveals several additional damage sites. This most probably results from much higher lateral resolution of SAM.

The sample with SNAP pretreatment shown in Figure 7 shows indication of severe localized damage. However, SAM is not able to markedly distinguish this corrosion site (also identified by SVET) from others observed on SAM scan. These results indicate that SNAP pretreatment definitely improves resistance to general corrosion. However, minor defects in SNAP leads to high localized corrosion activity that can result in failure.

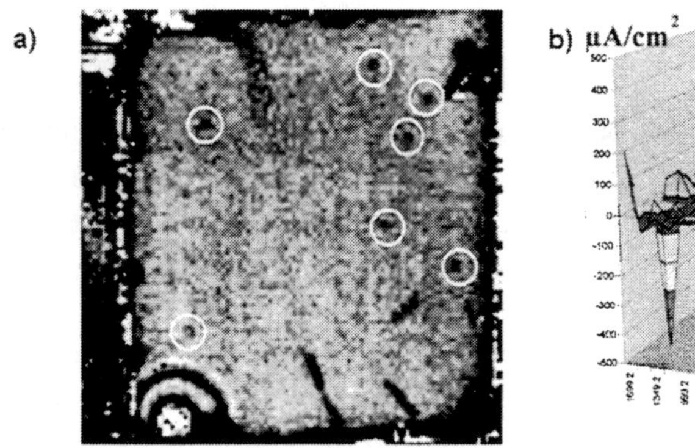


b) µA/cm$^2$

**Figure 7. 38 µm thick epoxy coating, <u>with SNAP layer</u>**
**a) SAM C-scan area: 6x6 mm$^2$, stepwidth: 50 µm , substrate reflection**
**White circles: hidden corrosion**
**b) SVET scan: 5x5 mm$^2$, 20x20 steps**

Figure 8 shows consecutive SVET, Fan Thermography and SAM images of exactly the same area of another coated sample. These scans were taken from an Al 2024-T3 panel coated with 50 µm thick proprietary epoxy coating. The sample was exposed for 125 hours at open circuit potential in Harrison's solution. SVET measurements were carried out first as discussed earlier. Figure 8a presents a vector overlay mapping of electrochemical current density superimposed over a video image. There are indications of delamination in the SVET video image. The longer vectors indicate increased corrosion activity underneath these delaminated regions.

Coating delamination or sites of corrosion are thermal barriers. If the test (object) surface is heated, the heat diffusion into the aluminum plate is inhibited in the delaminated or corroded region. This causes specific thermal response. The time for the maximum contrast at the surface is called response time. This time depends upon the thermal properties of coating and substrate and the coating thickness. Figure 8b shows a thermal contrast for corrosion below this coating. The spatial resolution was improved with a microscopic optic used in combination with the camera. The bright spots (higher surface temperature) indicate regions of reduced heat diffusion and correlate well with the SVET image.

The sample was then scanned with SAM. The substrate reflection C-scan is presented in Figure 8c. The SAM scan clearly identifies the same three spots observed by SVET and FAM (Figures 8a and 8b). At the same time some additional spots are visible. This could be due to the rather poor performance of this coating and the fact that repeated heating and additional exposure in a water bath has a deteriorating effect on the sample. The results presented demonstrate that the electrochemical and NDE techniques provide complimentary information. At the same time these techniques used together can provide detailed information about the state of the coating and there is a good possibility of integrating the information that can provide the details of the health of the substrate beneath a coating exposed to corroding environment [19].

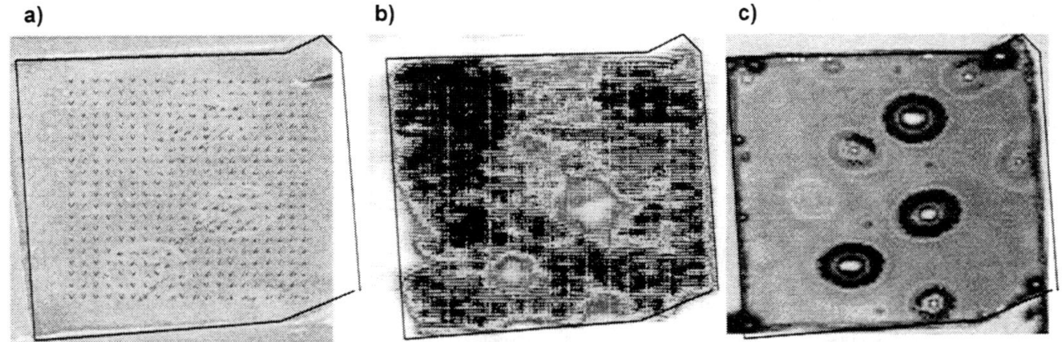


**Figure 8. NDE imaging of same sample: 50 μm thick artificially damaged epoxy coating [19]**
**a) SVET video image + current density vector overlay (vector area: 4x4 $mm^2$)**
**b) Fan Thermography image (image area: 6x6 $mm^2$)**
**c) SAM substrate reflection C-scan (scan area: 6x6 $mm^2$, stepwidth: 50 μm)**

## 4 CONCLUSIONS

The SVET was successfully used to monitor the initiation and progress of corrosion in coated panels exposed to Harrison's solution. An increase in current activity with exposure was noticed for poor coating systems over the exposure period. Little corrosion activity was noticed for panels protected with coatings currently used for U.S. Air Force aircraft. A variety of chemical, mechanical, and electrochemical techniques were investigated, and a relatively controlled method of creating artificial corrosion damage was established. It was possible to cause extensive corrosion damage to occur in just a few minutes by imposing external voltage.

SAM has proven to be an excellent laboratory technique. Both SAM and FAM were also successfully employed to detect sites of delamination and localized corrosion in aluminum alloy substrate under the coating. Reasonable correlation was obtained amongst SVET, SAM and Fan Thermography. The results from all the three techniques provided a complete detail of the damage state at the substrate under various coatings. Ongoing work is concentrated to distinguish corroded and delaminated regions.

## 5 ACKNOWLEDGMENTS


The work presented in this paper is sponsored by the Defense Advanced Projects Agency (DARPA) Multidisciplinary University Research Initiative (MURI), under Air Force Office of Scientific Research grant number F49620-96-1-0442. The authors would like to thank Ms. Angela Mahan, Dr. Andrew Vreugdenhil, and Dr. Joel Johnson, for their contributions and support.


## 6 REFERENCES


1. Khobaib, M., Final Report for US Air Force, Contract Number F 33615-94-C-58-04, (1997).
2. Hoffmann, J., et al., 3rd Annual Report for DARPA-MURI, Grant Number F49620-96-1-0442, (1999).
3. Jeffcoate, C.S. et al., in Proceedings of 43rd International SAMPE Symposium and Exhibition, Vol.43, (1998), pp. 2113-2122.
4. Taylor, S. R., "Assessing the Moisture Barrier Properties of Polymer coatings Using Electrical and Electrochemical Methods", IEEE Transactions on Electrical Insulation., Vol. 24, No. 5, Oct 1989
5. Mansfeld, F., "Analysis and Interpretation of EIS Data for Metals and Alloys", Issue AA: February 1993, Schlumberger Technologies, Billerica, MA.
6. Jeffcoate, C.S., Wocken, T. L., and Bierwagon, G. P., "Electrochemical Assassment of Spray-Applied Thermoplastic Coating Barrier Properties," J. Materials Eng. & Performance, 6 (1997) pp. 417-420.
7. Amirudin, A., and Thiery, D., "Application of Electrochemical Impedance Spectroscopy to Study the Degradation of Polymer-Coated Metals", Progress in Organic Coatings, 26, 1-28, (1995)
8. Macdonald, J. R., "Impedance Spectroscopy", Wiley, New York, 1987.
9. Isaacs, H. S., Aldykiewicz Jr., A. J., Thierry, D., and Simpson, T. C. "Measurements of Corrosion at Defects in Painted Zinc and Zinc Alloy Coated Steels Using Current Density Mapping", *Corrosion*, Vol. 52, No. 3, (1996) pp.163-168.

10. Applicable Electronics, Inc., 22 Buckingham Drive, Sandwich, MA 02563.
11. Aldykewicz Jr., A. J., Isaacs, H. S., and Davenport, A. J. "The Investigation of Cerium as a Cathodic Inhibitor for Aluminum-Copper Alloys", *Journal of the Electrochemical Society*, Vol. 142, No. 10, (1995) pp.3342-3350.
12. Isaacs, H. S., Davenport, A. J., and Shipley, A. "The Electrochemical Response of Steel to the Presence of Dissolved Cerium", *Journal of the Electrochemical Society*, Vol. 138, No. 2 (1991) p.391.
13. Kinra, V.K., and Iyer, V.R., Ultrasonics, Vol.33, (1995), pp. 95-109.
14. Briggs, A., Acoustic Microscopy, Oxford University Press, New York, (1992).
15. Netzelmann, U., and G. Walle, Photoacoustic and Photothermal Phenomena, 10th International Conference, Rome, Italy, 1998, published in AIP Conference Proceedings, Vol.463, American Institute of Physics, Woodbury, New York, (1999), pp. 401-403.
16. Khobaib, M., Rensi, A., Matikas, T., and Donley, M.S., "Real Time Mapping of Corrosion Activity Under Coatings", In Press, Progress in Organic Coatings, 2001
17. Vreugdenhil, A., et al., to be published in Polymeric Materials: Science & Engineering, Vol.85, (2001).
18. Hoffmann, J., Sathish, S., and Meyendorf, N., "Evaluation of Aluminum Alloy Airframe Structures under Protective Coatings using Acoustics and Thermographic Techniques", Proceedings of 2nd International Symposium for Transportation Systems, May, 2001, Pusan, South Korea
19. Hoffmann, J., Sathish, S., and Khobaib, M., "Acoustic and Thermographic Evaluation of Corrosion Protective Polymeric Coatings on Aluminum Alloy Airframe Structures", Proceedings of SPIE's 6th Annual Symposium on Nondestructive Evaluation and Health Monitoring of Aging Infrastructure, March, 2001, Los Angeles, CA

# AUTHOR INDEX

## NDE-Vol. 21
## NDE Challenges of the 21st Century: Theory to Practice


Avdelidis, Nicolas P. .... 79
Borinski, Jason .... 91
Brar, Amarjit S. .... 57
Cassino, Christopher .... 91
Cetinkaya, Cetin .... 71
Chen, Chi-Hau .... 95
Desai, Chandrakant S. .... 9
Donley, Michael S. .... 127, 141
Druffner, Carl .... 57
Duke, John .... 91
Ehsani, M. .... 29
Ghonem, H. .... 49
Herzog, Pamela G. .... 119
Hoffmann, Jochen .... 83, 141
Huang, J. .... 135
Inoue, Akihiro .... 21
Keller, Michael S. .... 9
Khobaib, Mohammad .... 127, 141
Koui, Maria .... 79
Kundu, Tribikram .... 9, 29, 37
Kwon, Jeong-Rock .... 113
Li, Chen .... 71
Liaw, Benjamin .... 65
Lin, Jiadao .... 71
Liu, Yangxiong .... 65
Lupien, Vincent .... 119
Maruyama, Shinichi .... 21
Meyendorf, Norbert G. .... 83
Miller, James T. .... 119
Moles, Michael .... 119
Moropoulou, Antonia .... 79
Na, Won-Bae .... 37
Palakal, M. J. .... 135
Peeler, D. T. .... 135
Peterson, Michael .... 105
Pidaparti, R. M. .... 135
Pierides, Alexis M. .... 65
Qidwai, Uvais .... 95
Raikar, Ganesh N. .... 57
Rebbapragada, S. .... 135
Rösner, Henrik .... 83
Sathish, Shamachary .... 57, 141
Schumaker, Edward J. .... 57
Selman, John J. .... 119
Shen, Liming .... 57, 83
Shin, Hyeon Jae .... 113
Song, Sung-Jin .... 113
Sotiropoulos, D. A. .... 1
Sugiura, Toshihiko .... 21
Sun, Miao .... 105
Towfighi, S. .... 29
Yoshizawa, Masatsugu .... 21
Zaki, A. S. .... 49


**Book Number: I00551**